Artisan/Practitioners and the Rise of the New Sciences,
1400–1600

The OSU Press Horning Visiting Scholars
Publication Series

EDITORS: *Anita Guerrini and David Luft*

Artisan/Practitioners and the Rise of the New Sciences, 1400–1600

PAMELA O. LONG

Oregon State University Press | CORVALLIS

Cover art: Vitruvius, *De architectura libri dece . . . da Cesare Caesariano* [Como]:
G. da Ponte, [1521], fol. 165r. Courtesy Rosenwald Collection, Library of Congress,
Washington, D.C.

The paper in this book meets the guidelines for permanence and durability of the
Committee on Production Guidelines for Book Longevity of the Council on Library
Resources and the minimum requirements of the American National Standard for
Permanence of Paper for Printed Library Materials Z39.48-1984.

Library of Congress Cataloging-in-Publication Data
Long, Pamela O.
  Artisan/practitioners and the rise of the new sciences, 1400-1600 / Pamela O. Long.
    p. cm. -- (Horning visiting scholar lecture series)
  Includes bibliographical references and index.
  ISBN 978-0-87071-609-6 (pbk.)
  1. Science, Medieval. 2. Science--Europe--History. I. Title.
  Q124.97.L66 2011
  509'.40902--dc23

                         2011019529

© 2011 Pamela O. Long
Second printing 2014
Printed in the United States of America

Oregon State University Press
121 The Valley Library
Corvallis OR 97331-4501
541-737-3166 • fax 541-737-3170
http://oregonstate.edu/dept/press

For Bob Korn

Allison Rachel Korn

Marco Yunga

and Lucas Samay Yunga Korn

Contents

Illustrations

FOREWORD

The Thomas Hart and Mary Jones Horning Endowment in the Humanities at Oregon State University was established by a bequest from Benjamin Horning (1890–1991) in memory of his parents, Mary Jones and Thomas Hart Horning, members of pioneering families of Benton County and Corvallis, Oregon. Benjamin Graham Horning graduated from what was then Oregon Agricultural College in 1914, and went on to complete a medical degree at Harvard and a degree in public health from the Johns Hopkins University. Dr. Horning's long professional career included service in public health in Connecticut, work on rural health as a staff member with the American Public Health Association, and a position as medical director for the W. K. Kellogg Foundation, which led to his spending many years in Latin America. Dr. Horning wanted his bequest at Oregon State University to expand education in the humanities and to build a bridge between the humanities and the sciences.

Since 1994, the endowment has supported an annual lecture series and individual lectures, conferences, symposia, and colloquia, as well as teaching, research, and program and collection development. The Horning professorships are housed in the Department of History. The first Thomas Hart and Mary Jones Horning Professors in the Humanities, Mary Jo Nye and Robert A. Nye, were appointed in 1994. Anita Guerrini and David A. Luft succeeded them in 2008. The Horning Visiting Scholar in the Humanities program was inaugurated in 2006 to allow a distinguished scholar to spend a week in residence at OSU and deliver a series of lectures as well as participate in other activities in and out of the classroom. Visiting Scholars since 2006 have included Ken Alder (Northwestern University), Liba Taub (Cambridge University), Lawrence Principe (Johns Hopkins University), and John Beatty (University of British Columbia).

The OSU Press Horning Visiting Scholars Publication Series, under the direction of the Press's acquisitions editor, Mary Elizabeth Braun, publishes the public lectures delivered by the Horning Visiting Scholar: one volume in the series has appeared, Liba Taub's *Aetna and the Moon* (2008). Other works in the humanities outside the scope of this series that the series editors, Anita Guerrini and David Luft, have found to be relevant to the aims of the Horning Endowment may also be published by the Press in the future.

Pamela O. Long was the Horning Visiting Scholar in April 2010. Dr. Long is an independent scholar of late medieval and Renaissance science and technology based in Washington, D.C. Her research and scholarship have focused particularly on craft traditions, authorship, the intersections of architecture and science, and the history of engineering. Dr. Long's scholarship has ranged across Europe, with a recent focus on the city of Rome. Her many publications include *Openness, Secrecy, Authorship: Technical Arts and the Culture of Knowledge from Antiquity to the Renaissance* (2001), which won the 2001 Morris D. Forkosch Prize for the best book in intellectual history, awarded by the *Journal of the History of Ideas*; *Obelisk: A History* (with Brian Curran, Anthony Grafton, and Benjamin Weiss, 2009); and the three-volume *Book of Michael of Rhodes: A Fifteenth-Century Maritime Manuscript* (2009), co-edited and co-authored with David McGee and Alan M. Stahl. She co-edits, with Robert C. Post, the booklet series *Historical Perspectives on Technology, Society and Culture*, co-sponsored by the American Historical Association and the Society for the History of Technology.

In *Artisan/Practitioners and the Rise of the New Sciences, 1400–1600*, Pamela Long offers a concise and compelling account of the roles of artisans and practitioners in the development of the new sciences of the early modern era. Most accounts of the "scientific revolution" have emphasized elite academic natural philosophers, and the role played by craftspeople has been hotly contested among historians. Intriguingly, Dr. Long begins her book with a reassessment of earlier historiographical accounts of this issue, focusing particularly on the Marxist philosopher Edgar Zilsel (1891–1944). Subsequent chapters draw on Dr. Long's deep knowledge of learned and craft traditions to argue that, in fact, these traditions found increasingly common ground over the course of the fifteenth and sixteenth centuries to create the empirically based new sciences. The humanist revival of the works of the ancient Roman architectural scholar Vitruvius, for example, led engineers, craftsmen, and scholars to collaborate in construction and engineering projects. Dr. Long effectively employs the concept of "trading zones" (developed by the historian of science Peter Galison to talk about modern physics) to talk about places ranging from mines to cities where interchanges between scholars and artisans led to the creation of new knowledge. Her book adds a substantial new dimension to ongoing debates on the origins of modern science.

ANITA GUERRINI

PREFACE

This book came out of the public lectures that I gave as the Horning Visiting Scholar at Oregon State University in April 2010. Three of the essays in this book are a direct result of the lectures, while a fourth, on the Vitruvian tradition, is an addition. It was a great pleasure to spend the week at Oregon State to give lectures to wonderful, inquisitive audiences, to talk extensively throughout the week with faculty and graduate students in the history department and in other departments of the university, and to participate in seminars and classes. I especially thank Horning Professors Anita Guerrini and David Luft, who invited me, warmly welcomed me, and made my stay a memorable one. I thank Lisa Sarasohn and Jacob Hamblin for inviting me to participate in their classes and seminars, Michael Osborne and Anita Guerrini for their welcoming hospitality, and Elissa Curcio for making and facilitating numerous arrangements.

One might think that delivering a series of three lectures would be a simple matter of organizing what one knows and delivering that knowledge in a rhetorically effective way to a (hopefully) receptive audience. This was hardly the case with these lectures on artisan/practitioners and their influence on the new sciences. It is true that I have been thinking about and investigating premodern artisans—people whose work life was centered on the skilled manipulation of materials in order to fabricate objects, and those who engaged in complex material practices such as agriculture or navigation—and artisanal culture for most of my scholarly life. The intense months of preparation for the lectures, however, involved new primary research and the assessment of much recent scholarship on relevant topics. Preparing for the lectures also provided an unexpected opportunity to step back and to reassess my own previous work and thinking on the topic and to create a new synthesis that went beyond that work.

During the months of transforming the lectures into chapters of this book, I was supported by National Science Foundation grant #0849158. The entire manuscript was critically read by Horning Professor Emerita Mary Jo Nye; Anita Guerrini; my sister, master writer and poet Priscilla Long; and astute reader Bob Korn. Together they saved me from errors and greatly helped improve the work. Finally, I thank Teresa Jesionowski for superb copyediting that saved me from many errors, and I thank the acquisitions editor of Oregon State University Press, Mary Elizabeth Braun, and managing editor, Jo Alexander, who brought the book to press with efficiency and care.

Artisanal Values and the Investigation of Nature

A long-standing issue in the history of science can be framed as a question: Did artisan/practitioners influence the development of the new sciences? By the expression "artisan/practitioners," I refer to a broadly diverse group of skilled artisans such as weavers and instrument makers; architect/engineers involved in the design and construction of buildings, bridges, and the like; and practitioners such as farmers and navigators.[1] Through the lens of this one issue, which is a fundamentally important one, this book treats a series of complex, multifaceted, and long-term developments, traditionally referred to as the "scientific revolution." The people contributing to this development did not use the phrase "scientific revolution" but often referred to their own work as a "new science." For example, Galileo Galilei (1564–1642) called one of his most important books *Discourse on the Two New Sciences* (referring to the strength of materials and the motion of objects). It is this term, commonly used in the sixteenth and seventeenth centuries, that I use throughout.

The new sciences are usually considered to have developed on the European continent and in England from the mid-sixteenth century through the seventeenth century. A key initial event was the publication in 1543 of the *De revolutionibus* by Nicholas Copernicus (1473–1543), and a concluding highpoint is often considered to be the publication of the *Principia Mathematica* (1687) by Sir Isaac Newton (1643–1727). This book begins earlier, around 1400, and ends around 1600. I argue that in this two-hundred-year period, the empirical values that were intrinsic to artisanal work came to be embedded within a broader European culture. Near the end of this period, around the mid-sixteenth century, new ideas about the cosmos and the natural world began to emerge. New kinds of questions began to be asked, and new methods began to be utilized to investigate nature.[2]

Changes brought by the new sciences must be understood against a background of the Aristotelianism and scholasticism of the medieval universities. The *studium generale*, as the medieval university was called, emerged in the early thirteenth century. Professors in the universities taught about the cosmos and the natural world, or "natural philosophy," by focusing on Aristotle and Aristotelian texts and commentaries. Such texts formed a rich tradition to which there was no equivalent in terms of depth, breadth, or sophistication. It included time-tested views concerning the structure of an Earth-centered cosmos, as well as a material and causal understanding of substances, and a coherent explanation of motion and change. In addition, this tradition contained within it a set of methodologies or ways of approaching the investigation of the natural world.[3]

Substantial scholarship underscores the depth and richness of this Aristotelian natural philosophy, as well as its ability to develop and change. It was an important part of the educational system until at least the end of the seventeenth century.[4] At the same time, though, from about the mid-sixteenth century, new ideas and approaches concerning the structure of the cosmos and concerning phenomena such as motion and substance began to be put forward. Innovative approaches and new instruments encouraged the practice of observation—observation of plants and animals, stars and planets. It is a distortion, however, to say that Aristotelianism was replaced by the new sciences in this era; rather, they coexisted, sometimes in fruitful dialogue and synchrony, sometimes in polemical opposition.

Changing views about how the world was constructed and how it should be investigated came in the wake of social, economic, and political changes that had occurred in medieval and late medieval Europe. The rise of the cities; the development of commercial capitalism and long-distance trade; the development of large-scale industries such as textiles, armaments, and mining; the immense expansion of overseas markets;[5] oceanic explorations and the discovery of lands new to Europeans; new knowledge about never-before-heard-of flora and fauna and unknown peoples;[6] the increasing political and social importance of princely courts;[7] the invention of movable type and printing and the subsequent great expansion of books and pamphlets of all kinds;[8] conspicuous consumption among elites and the parallel expanding manufacture of luxury goods; the construction of massive palaces and other buildings and their ornamentation;[9] the increased importance of visual culture and the rising status of the visual arts aided by the invention of artists' perspective[10]—all this brought with it a growing valuation of things, of objects, and often, an appreciation of the skill and knowledge of the people who made those objects. These closely interrelated complex historical developments influenced the ways in which the world was approached, investigated, and understood.

The new sciences put forth innovative ideas about the structure of the cosmos and other aspects of the natural world. For example, in his *De revolutionibus*, Copernicus rejected the geocentric model of the cosmos and proposed a heliocentric model. Tycho Brahe (1546–1601) proposed a third, compromise model, a geoheliocentric cosmos in which the moon and the sun revolve around the earth, while Mercury, Venus, Mars, Jupiter, and Saturn revolve around the sun, and the fixed stars around all.[11] New approaches were also adopted in other areas such as anatomy, physics, and natural history.

The new sciences entailed particular discoveries and new ideas, and, in addition, changes in the kinds of questions asked and in the methods used to answer those questions. Investigators increasingly adopted a variety of empirical approaches and values as relevant to knowledge about the world—an appreciation for the knowledge acquired by hands-on manipulation and the use of instruments; the practices of direct observation and experimentation; methods of precise measurement and other forms of quantification; and a positive valuation of individual experience. These values and practices closely resembled those held by contemporary artisans and practitioners such as painters, sculptors, carpenters, weavers, potters, architect/engineers, mariners, apothecaries, and farmers.[12]

In the 1920s, as the modern discipline of the history of science developed, this sixteenth-century overlap and coincidence of the values and practices of artisans, on the one hand, and investigators of the natural world, on the other, suggested to some that artisans had significantly influenced the development of the scientific revolution. Other historians vociferously disagreed. Since the 1980s, the issue has reemerged but framed in a very different way. The history of this issue and the issue itself are the twin subjects of this book.

"Artisan/practitioners" were men and women who worked with their hands in craft production (for example, carpenters, weavers, and instrument makers) or carried out complex practical tasks such as farming or navigation. A characteristic of the medieval period in general is that the world of crafts and the world of learning existed in quite separate realms. The study of the natural world occurred in the universities and was called natural philosophy. Craft practice involved a hands-on process in which apprentices learned by doing and making, sometimes formally under an apprenticeship contract regulated by a guild and sometimes informally as part of a family unit. Crafts such as spinning, weaving, and painting and practices such as agriculture and navigation were learned under the guidance of a skilled practitioner, often an older family member. In some cases writings surrounded craft activities, such as records and accounts kept, regulations for the craft, and specifications created by patrons for particular works, and there may have been other kinds of communicative and mnemonic devices such as drawings on paper or models made out of wood. The usual activity of craft practice, however, was carried out by making something or carrying out some physical task. Any instruction was usually oral and was provided in one of the vernacular languages, whether Italian, French, German, English, or some other language.[13]

In contrast, reading, writing, and teaching at the universities in the Middle Ages were conducted in Latin. University instruction in natural philosophy as well as in other topics was based on two main activities, the lecture (*lectio*) and the disputation (*disputatio*). In the *lectio*, the professor lectured and commented on authoritative books that were prescribed by regulations that governed the university. These books included ancient texts such as those by Aristotle, sometimes with additions and explications by medieval commentators. Edward Grant describes the ways in which the lecture developed over several centuries—it consisted of the summary of one or more texts, and a discussion or commentary, sometimes including reference to previous commentaries. The disputation, or *disputatio,* took several forms but always involved one or more questions (the *quaestio*). Often students under examination would be given opposite sides of a question to defend. The resolution of the question, called the determination, or *determinatio,* would be summarized by the presiding master.[14]

Impinging on this world of university scholasticism, a new intellectual movement that came to be known as humanism arose in the fourteenth century—most notably in the writings of Petrarch (1304–1374) and then in those of Coluccio Salutati (1331–1406). Humanism became highly influential by the early fifteenth century. Humanists wanted to reform the Latin language by returning to the Latin usage of classical antiquity, as exemplified in the writings of Cicero (106 BCE–43 BCE), and by expunging what they considered to be crude medieval corruptions. Humanists also turned from the scholastic interest in logic, philosophy, and theology to rhetoric, moral philosophy, and history. The humanists admired and investigated the ancient past. They searched for new texts and studied and reedited them. They also investigated ancient artifacts and ruins, searched for ancient sculptures, and collected ancient coins and medals. Many humanists earned their livelihoods by serving as secretaries to princes, popes, cardinals, and other elite men, using their well-honed Latin skills in the process. They created an intellectual and bookish world concerned with a variety of skilled practices and physical objects. Humanism initially grew up outside of the universities, but eventually, by the late fifteenth century, it began to influence the course of studies within them.[15]

The new culture of humanism had important implications for the artisanal crafts. Humanism emphasized practical life as opposed to scholastic logic, and it encouraged an interest in material goods and in the decorative arts that were part of daily life. It is notable that one of the greatest humanists of the fifteenth century, Leon Battista Alberti, who was educated at a university and possessed highly developed Latin literary skills, also wrote about sculpture, painting, and architecture, and practiced painting and architecture. Some humanists like Alberti wrote on the practical and technical arts. Practitioners themselves also began to write books with increasing frequency about their own disciplines. Writings about various practices and artisanal crafts found a readership among patrons and others in the elite and learned classes as well as among other practitioners.[16] Although writings on technical arts and practices had occurred since antiquity, such writings expanded rapidly in the fifteenth and sixteenth centuries. This proliferation of writings made the values of artisanal culture more readily available to be used as components of methodologies for the investigation of nature.

This book is aimed at a nonspecialist readership, and it is hoped that it may be of some use to historians of science and technology as well. Chapter 1 treats the history of the idea of artisanal influence on the new sciences as that idea emerged in the 1920s and 1930s. Such notions developed for the most part (but not exclusively) within the context of Marxism and Marxist notions concerning capitalist development. A second important context was the Vienna Circle in the 1920s and 1930s and the logical empiricism or logical positivism with which it was associated. Edgar Zilsel (1891–1944) developed the "Zilsel thesis" of artisanal influence on the scientific revolution. Zilsel lived and worked in Vienna and took part in the meetings of the Vienna Circle before he emigrated to the United States in 1939, where he wrote his seminal articles. In another context, the Russian physicist Boris Hessen (1893–1936) read a famous paper relevant to the issue of artisanal influence in London in 1931 at the Second International Conference for the History of Science and Technology. Shortly thereafter, in Frankfurt, Germany, a controversy broke out over the origins of the scientific revolution, both sides of which invoked aspects of the artisanal world. The protagonists were Henryk Grossmann (1881–1950) and Franz Borkenau

(1900–1957), both at the Institute for Social Research, known as the Frankfurt School.

Chapter 1 also discusses the views of non-Marxist scholars who address, in one way or another, the issue of artisanal influence on the new sciences. One is the philologist Leonardo Olschki (1885–1962), who in the 1920s wrote a three-volume work on technical writings in the fifteenth and sixteenth centuries and their importance for Galileo. Another is the American sociologist Robert Merton (1910–2003). The chapter also treats opposition to the view of artisanal influence—a position that prevailed among Anglo-American historians of science after the Second World War. It concludes with a discussion of the reemergence of the issue in recent decades within frameworks quite different from the original Marxist one.

In the remaining three chapters I explore the issue of the influence of artisans on the new sciences in the two-hundred-year period from 1400 to 1600 on the basis of primary source materials and from my own point of view. As will become apparent in these chapters, in general I think that artisanal influence on the new sciences was significant, but I suggest that the dichotomous categories with which the issue traditionally has been discussed— artisan/scholar, handworker/theorist, practice/theory, experimental/mathematical, art/nature—represent distorting lenses. It was precisely the blurring of these traditionally separate categories that provided the modality for the influence of artisanal values on the new sciences. Some artisans took up pens and began to write books, while some learned men began to take up artisanal practices such as surveying and measurement. Further, the distinction between artifactual objects, that is objects made by humans, and natural objects came to be blurred— for example, the potter Bernard Palissy (ca 1510-1589) fabricated platters embedded with lizards.

Chapter 2, on "art" (i.e., artisanal crafts) and "nature," discusses the historical interaction between these two changing concepts. This chapter first centers on the Aristotelian view of art and nature. It then turns to the question of the role played by experience and experiment in medieval natural philosophy and in medieval alchemy. The degree to which experiment and empirical manipulations (that is, art) could lead to an understanding of nature depended on the assumed relationship between art and nature. By the late fifteenth century, there was a growing interchangeability of the two categories, and a growing tendency to use machines and instruments to investigate and discuss natural phenomena such as power and motion.

Chapter 3 treats the Roman architect Vitruvius (fl. 40s–20s BCE) and the Vitruvian tradition as an important common ground on which practitioners and university-educated men came together to discuss substantive issues. Vitruvius's treatise, the *De architectura,* was the only architectural treatise to survive intact from the ancient world. Beginning in the fifteenth century, a rich tradition of architectural practices and writings, including commentaries on the *De architectura,* picked up and developed the Vitruvian dictum that architecture consists of both fabrication and reason. Workshop-trained practitioners as well as university-trained humanists contributed to this written tradition, as they examined buildings and artifacts with the ancient text and measuring rods in hand. I suggest that Vitruvianism served as a modality for empirical investigation of issues involving building construction, hydraulics, and machines. Problems in understanding the *De architectura* in view of extant ancient buildings or ruins led to significant communication between practitioners and the learned. The Vitruvian tradition became a "trading zone" in which substantive communication occurred between the two groups.

The final chapter focuses on other "trading zones"—arenas in which the unskilled learned and skilled practitioners exchanged substantive knowledge. In the late sixteenth century, numerous locales served this purpose. These included arsenals, mines, workshops, and cities. Such places became important sites for communication and exchange between men trained as artisans and those schooled in Latin learning. Men from these diverse backgrounds exchanged information concerning material production and problems in engineering, but also concerning the nature of materials and of natural phenomena—traditionally topics belonging to natural philosophy.

This book shows not necessarily a direct influence of particular artisans on specific individuals investigating the natural world, although such influence did occur. Rather, it shows that the categories of "art" and "nature" became less and less apropos as categories indicating separate entities as the two came closer together and even became interchangeable. At the same time, the categories "scholar" and "craftsman" as classifications of types of individuals became an oversimplification when in some arenas the two moved closer together, communicated, and adopted each other's practices. I suggest that empirical values, once held predominantly by artisan/practitioners, came to be generally adopted by the society at large, thereby making them readily available for use by investigators of the natural world. I further suggest that this development came about, at least in part, by the widespread development of "trading zones" in which the learned and the skilled communicated, exchanging substantive information. I suggest finally that this development cannot be attributed to artisan-trained individuals alone but rather came about through the interaction of artisanal and humanist culture.

CHAPTER I

Artisan/Practitioners as an Issue in the History of Science

In the 1920s and 1930s, just as the history of science as a discipline was taking shape, a new thesis emerged concerning the influence of artisans and artisanal culture on what was termed the "scientific revolution" of the seventeenth century. A group of scholars began to discuss the ways in which the mechanical arts—that is, the arts and crafts carried out by skilled artisans—influenced the development of the mechanical world view that emerged in the seventeenth century. The "mechanical world view" was shorthand for the idea that all motion and change was mechanical and that the universe itself functioned as a machine. The mechanical world view developed along with a complex of other ideas about the natural world and how to study it that are often grouped together under the term the "new sciences."

Four of the scholars who debated the thesis of artisanal influence adhered to one or another form of Marxism, and at least one, the Viennese physicist and philosopher Edgar Zilsel, was associated with the philosophical outlook known as logical empiricism. In addition to Zilsel, the Marxist scholars who developed versions of the thesis of artisanal influence were the Russian physicist Boris Hessen, the Viennese sociologist Franz Borkenau, and the Polish political economist Henryk Grossmann. All four were from Jewish backgrounds, and all suffered from the virulent anti-Semitism of their day. All were leftists who engaged extensively in both philosophical and political struggles. Their differing views on artisanal influence were tied to their larger philosophic and scientific outlook and to their political activities.[1]

Other scholars who were not Marxists also developed ideas about artisanal culture and the new sciences. One was the German-Italian philologist Leonardo Olschki, whose three-volume study of fifteenth- and sixteenth-century technical writings and their influence on Galileo, published in the 1920s, seems to have influenced at least some of the Marxists. A second, the American sociologist Robert Merton, wrote a dissertation the second half of which was devoted to technical and practical arts and their influence on the sciences. It was published in 1938 as *Science, Technology, and Society in Seventeenth-Century England.*[2]

This early work of scholars in the 1920s and 1930s focused on artisanal influence on the new sciences is worth revisiting because these discussions reflect significant interpretive issues in the discipline of the history of science as a whole. In addition, the assumptions of those decades have at times silently shaped current discussions in ways that would benefit from explicit analysis.

The Marxist Tradition

Several key concepts in the writings of Karl Marx (1818–1885) are relevant to this twentieth-century scholarship. Marx along with Friedrich Engels (1820–1895) had developed the tenets of historical materialism in the 1840s in opposition to the prevailing notions of idealism. Whereas idealists argued that ideas and beliefs were the moving forces of history, Marx suggested, instead, that history was driven by economic production. "Productive forces" included both the means of production, such as tools, machines, and factories, and labor power, which involved human skill, knowledge, and experience. Marx argued that the foundation of society was its economic structure, by which he meant the relations of production. All the rest—law, politics, social consciousness, intellectual life, and science—he considered to be superstructure determined in perhaps complex ways by the underlying structure of economic production.[3]

Individuals and political groups modified Marx's influential ideas in various ways during the first three decades of the twentieth century. Max Adler (1873–1937), an Austrian politician and social philosopher, who was an important leader of the Austrian Social Democratic Party, developed a strand of Marxism referred to, in Edgar Zilsel's Vienna, as Austro-Marxism. Adler conceived of Marxism as "a system of sociological knowledge" and argued that the economic determinism of Marxism should not be thought of as a materialist determinism; rather, economic production was mediated by consciousness. So, in a sense Adler took the materialism out of Marxism and suggested that even economic phenomena possessed a mental character. He transformed Marx's idea of economic production, making it not the material production of goods per se, but a category of knowledge, a concept that originated in reason and was not derived from experience. Adler also analyzed changing class structure and the process of differentiation within social classes. For example, he divided the proletariat into subclasses, including a labor aristocracy of skilled workers, an idea that may well have influenced Zilsel.[4]

Edgar Zilsel, undoubtedly the most prominent proponent of the thesis of artisanal influence, was a physicist and scholar who was born in Vienna in 1891 and studied mathematics and physics at the University of Vienna. Deeply influenced by Adler's Marxism, he joined the Austrian socialist party in 1918. He was active on the periphery of the Vienna Circle in the 1920s.[5]

The circle began as an informal group of philosophers and scientists that met regularly to discuss philosophical issues pertaining to science. Their discussions and the writings that emerged from them form an important grounding on which the philosophy of science as a discipline developed over the twentieth century. Their beliefs and ideas, taken together, are traditionally referred to as "logical positivism," or "logical empiricism." Although the developments and strands of this philosophical movement are outside the scope of this book, their views (which were by no means always in agreement) can be summarized by two important ideas. The first was that knowledge came only from experience. The second was that the task of philosophy was to clarify this experience by logical analysis. Vienna Circle philosophers were opposed to the idea that theory and metaphysics could be useful for philosophic and scientific knowledge.[6]

As a positivist, Zilsel believed that metaphysics should be abandoned, that the laws of social sciences and history could be formulated according to scientific principles. He also believed that the laws of history, like the laws of physics, could be discovered by empirical investigation. In the summer of 1923, he submitted his *Habilitation*-dissertation (the postdoctoral requirement for teaching at the university level) to the University of Vienna on the concept of genius; he was asked to withdraw it on the grounds that the topic was not sufficiently philosophical. When asked to submit a different work, he refused, saying that he did not want his research to be determined by external considerations. Without a *Habilitation*, he was unable to teach in the university, but he became deeply involved in the Viennese Institutes of Adult Education or *Volkhochschulen*—People's High Schools (which were adult education centers, rather than institutions similar to, for example, American high schools). He taught mathematics, physics, and philosophy between 1921 and 1934, at which time the newly established fascist dictator, Engelbert Dollfuss (1892–1934), abolished the institutes. In the following years, Zilsel taught physics in a Viennese *Gymnasium,* or high school.[7]

Zilsel's philosophy was shaped by his work in the Institutes of Adult Education. Along with Otto Neurath (1882–1945), a sociologist and a founder of the Vienna Circle, Zilsel believed that knowledge, life, and education should be governed by a principle of unity. Education should not alienate workers from their own cultural and social roots, but should rather create an expanded sense of unity in which knowledge was relevant to the concerns of daily life.[8] The unity of knowledge was not just a social ideal but a philosophical one, and here Zilsel disagreed with Neurath and others concerning how the unification of science should be reconstructed. For him the unity of the sciences was an empirical issue to be investigated, not an unquestioned assumption.[9]

In the 1920s, Zilsel expanded the material of his failed *Habilitation* into a book on the origin of the concept of genius. In it, he described the rise of the idea of genius and of individualism as an aspect of early capitalism and the competitive mentality associated with it. He also noted the emergence of the idea of the inventor, the discoverer, and the

1.1. Edgar Zilsel by unknown artist in Austrian expressionist style, painted during World War I, when Zilsel was in his early twenties. Photograph courtesy of Joanna Zilsel.

artist in the early Renaissance. Zilsel concluded his book with a section titled "Laws on the Concept of Genius." These were hypotheses that were meant to be tested by comparative studies with other cultures to determine under what historical conditions the idea of genius might arise. As Diederick Raven has shown, Zilsel's historical study was an aspect of his work on the unity of knowledge. He believed that history and humanities possessed laws similar to natural laws. Such laws must be discovered empirically, just as were the laws of nature. Zilsel's empirically oriented view of the unity of knowledge motivated his later work on the social origins of the scientific revolution.[10]

The Vienna Circle dissolved in the late 1930s as its members fled the Nazis. Edgar Zilsel and his family—his wife, Ella Zilsel, who had taught English and German at the women's *Gymnasium* in Vienna, and their sixteen-year-old son, Paul—were forced into exile after the *Anschluss* on March 13, 1938, that unified Germany and Austria under the Nazis. The family moved to New York via London. In New York

City in the early 1940s, while struggling to gain a foothold, Zilsel wrote several important articles on the contribution of what he called "superior artisans" to the development of the empirical sciences. In August 1943 he was offered a position at Mills College, a small women's college in Oakland, California. He moved to take up his position without his wife, who was mentally ill and chose to stay in New York, and without his son, who was in graduate school at the University of Wisconsin.[11] He had planned to continue his work on artisanal influence and to write a book on the subject. Instead, he tragically committed suicide in his office on March 11, 1944, six years after the Austrian *Anschluss* with Hitler's Germany. Zilsel was a logical positivist, but also a Marxist. He was one of a number of Marxist historians who in the 1930s developed ideas about the relationship of artisan/practitioners and the scientific revolution.[12]

In the early 1930s, other Marxist thinkers were developing ideas about the ways that artisanal culture contributed to the development of the scientific revolution. A key event for the expression of such a view was a meeting that took place in London in July 1931—the Second International Congress of the History of Science and Technology. This congress was attended by a Russian delegation formally led by the Marxist theoretician and Soviet politician Nicholai Bukharin (1888–1938). The delegation, which created a great stir in London, included a physicist, Boris Hessen, who was prominent in official Soviet scientific circles. Loren Graham notes that Hessen had been censored at home because he championed relativity theory, which the authorities took to be a form of bourgeois physics. The delegation included Arnošt Kolman (1892–1979), the party secretary, who had been instructed to report back concerning the performance of Hessen and Bukharin (who was also under suspicion).[13]

In his talk at the congress, titled "The Social and Economic Roots of Newton's *Principia*," Hessen expounded Marxist ideology far more effusively than he had in his talks in the Soviet Union. He also connected Newton's ideas in the *Principia* with the technical aspects of material production in the transition from feudalism to capitalism. Graham observes that everyone in the Soviet Union accepted Newtonian physics. If Hessen could show that Newtonian physics could be valued independently from the economic order from which it arose and from Newton's religious and philosophical conclusions, by extension, so also could relativity theory be separated from its "bourgeois" context of origin and surrounding philosophies. Whatever Hessen's success in London from the Soviet point of view, it was temporary. Both Bukharin and Hessen would become victims of the Stalinist purges. Hessen was arrested in Russia in August 1936, tortured seventeen times, tried for terrorism on December 20, 1936, and that day executed by a firing squad. Bukharin was executed in 1938.[14]

Before these grim events, in the paper presented at the congress, Hessen had rejected the idea that Newton's achievements were the result of individual genius. Instead, he endorsed Karl Marx's notion that changes in the economic production of commodities in turn produced changes in the superstructures of society. Hessen also adopted Marx's periodization. That is, he argued that a feudal economy existed in the medieval period and that it disintegrated as merchant capitalism and manufacturing arose in the sixteenth and seventeenth centuries. This rise brought about new demands for technology, especially in the areas of transportation, communication, artillery and other aspects of war, and mining. He viewed the development of natural science as a product of the technical needs of the new bourgeois class. He insisted that the core of Newton's *Principia* consisted of technical problems and that the themes of Newton's work, although not overtly technical, were determined by technical issues.[15]

Hessen suggested that classical mechanics—the Galilean and Newtonian mechanics that emerged in the seventeenth century—developed on the basis of the kinds of machines used in that century. It was relevant that industrial machinery such as lifting machines and waterwheels produced only mechanical motion. For him, the kinds of machinery available determined the nature of physics. Thus, thermodynamics could develop only after the eighteenth-century invention of the steam engine, and electromagnetism as a branch of physics could develop only after the nineteenth-century invention of electromagnetic machines. It followed that the mechanical philosophy that developed in the seventeenth century was a direct result of the mechanical motion of seventeenth-century machines.[16]

Shortly after Hessen delivered his paper in London, the Viennese sociologist Franz Borkenau published an article on the origins of the empirical methodologies in the new sciences. Borkenau was born in Vienna in 1900. In 1918 he attended the University of Leipzig and also became involved in the German communist party (which he left in 1929). In the mid-to-late 1920s, he became associated with the Institute for Social Research (*Institut für Sozialforschung*) at Frankfurt, the well-known institute that came to be known as the Frankfurt School. In 1932 Borkenau published a paper called "The Sociology of the Mechanistic World Picture," followed in 1934 by a book, *The Transition from the Feudal to the Bourgeois World-Picture,* which delineated the relationship of labor, particularly craft labor, to the scientific revolution. Borkenau's article and book were published under the auspices of the Institute for Social Research.[17]

The institute originated in 1922 at the University of Frankfurt at the behest of Felix Weil (1898–1975) with the help of his father, Hermann Weil (1868–1927), a wealthy merchant who had founded a grain company in Argentina. The company, Hermanos Weil, eventually controlled the Argentine grain trade and maintained offices in all major European cities with about three thousand employees and sixty ships operating under the company flag. Following the death from diabetes of the prospective first director of the institute, Kurt Gerlach (1886–1922), the Viennese professor Carl Grünberg (1861–1940) agreed to be the director. Grünberg, a Marxist who taught law and political science at the University of Vienna, had been the publisher of the *Archiv für die Geschichte des Socialismus und der Arbeiterbewegung* (known as the Grünberg Archives). As the director of the institute between 1924 and 1927, Grünberg oversaw the publication of the collected works of Marx and Engels.[18]

An important influence on the institute as a whole and on Franz Borkenau in particular was the Hungarian philosopher, literary critic, and political leader Georg Lukács (1885–1971) and his book *History and Class Consciousness,* first published in 1922. Lukács had rejected Karl Marx's view of the deterministic one-to-one relationship of modes of production to superstructures such as philosophy and science. He also opposed the fashioning of Marxism into a type of sociology à la Max Adler. Central to his philosophy was the concept of reification, an idea that originated in the writings of Marx. Reification occurs when human activity becomes alienated from the person and is turned into a thing or a commodity; it is objectified and thus becomes a nonhuman object. Thus a person's labor performed on a time clock to produce part of a product on a factory line is reified labor, a commodity without real connection to the working individual.[19]

Influenced by Lukács and the idea of reification, Borkenau provided a detailed analysis of the economic and social forms of feudalism, Renaissance early capitalism, and seventeenth-century manufacturing capitalism. He believed that a revolution in thought had occurred in the transition from feudalism to bourgeois capitalism. The result of that revolution was the view of the world held by Descartes and other philosophers of the seventeenth century—a world that could be described mathematically and worked like a machine. Borkenau here accepted Marx's view of an economic transition in which the craft production of the medieval workshop was transformed into other forms of manufacture. As he described it, craft workers were now gathered into one shop. Their labor had been reified into a commodity; they worked in a group in which each produced only a piece of the whole product because their jobs had been decomposed into parts. The worker was described only in terms of quantitative labor and pure physical movement. As labor became quantified and abstract, so also did the world—resulting in the mathematical/mechanistic world picture. Human labor became a commodity, an object, which was reinforced by a view of a world that functions entirely in a mechanical way, like a machine.[20]

Borkenau's thesis came under attack by another member of the Frankfurt Institute for Social Research, Henryk Grossmann. Grossmann was born in Krakow in 1881 to a family of well-to-do mineowners.

He studied jurisprudence, political economy, and economic history at the University of Krakow, was active in a radical socialist student organization, and worked intensely to revitalize the Jewish labor movement. Subsequently, he studied at the University of Vienna. In 1926 he became associated with the Institute for Social Research in Frankfurt as Carl Grünberg's assistant. It was the new director of the institute as of 1930, Max Horkheimer (1895–1973), who urged him to write a critique of Borkenau's traditional Marxist analysis. Horkheimer led a group of younger scholars who were disenchanted with the traditional Marxist leadership of the institute. Grossmann's critique of Borkenau can be seen as part of this younger-generation critique against the traditionalists.[21]

Grossmann took up Horkheimer's challenge and wrote an extensive critique of Borkenau's argument, which he took to be the notion that the mechanistic world view came directly out of the manufacturing process. He argued that his colleague had overgeneralized in his scheme of a three-stage development that proceeded from a feudal regime to a renaissance based on craft production to an early seventeenth-century manufacturing culture based on a factory production. Borkenau's view of history provided an excessively linear scheme of development, based as it was on a single world view for each period. Grossmann suggested, alternatively, greater recognition of overlap, differentiation, diversity of geographic area, and greater specificity of research into particulars. He viewed Borkenau's description of seventeenth-century manufacture as simply incorrect. The development of capitalist manufacture was far more complex, he thought, and it had begun much earlier than Borkenau would have it and, among other points, involved the use of skilled craft labor, not its abolition.[22]

Grossmann proposed an alternative to Borkenau's thesis that the mechanical world view developed out of the forms of manufacture. Instead, he argued that this world view came from the direct observation and use of machines and the changes in technological processes that those uses signaled. For him, "in the turning of the water wheels of a mill or of an iron mine, in the movement of the arms of a bellows, in the lifting of the stamps of an iron works, we see the simplest mechanical operations." He believed that the basic concepts of modern mechanics developed when men observed "the simple quantitative relations between the homogenous power of water-driven machines and their output." His case in point was Leonardo da Vinci (1452-1519), whose mechanical views and conceptions resulted from his experiences with the machine technology of his time, and whom Grossmann described as the founder of modern mechanics. He stressed that the technological developments of the fifteenth and sixteenth centuries in gunpowder, firearms, clocks, lifting mechanisms, and waterworks were key to the development of the mechanical philosophy that saw the world and everything in it operating like a machine.[23] Grossman's critique of Borkenau included ideas similar to those of Boris Hessen in his 1931 London paper, which had posited, as we have seen, the development of the mechanical philosophy out of mechanical machines. Grossmann was apparently unaware of Hessen's paper during his early dispute with Borkenau, although he later defended the Russian physicist's ideas.[24]

The years in which the Institute for Social Research published the papers of Grossmann and Borkenau in their *Zeitschrift* saw the rise of Nazi power. The institute, which possessed an endowment provided by Hermann Weil, had the foresight to transfer its money to Holland in 1931. Thus it was able to move from Germany after the Nazi assumption of power on January 30, 1933, to Geneva, Paris, London, and finally New York, where in 1934 it found a home at Columbia University (and remained there until its return to Frankfurt in 1950). After the institute under the direction of Horkheimer had settled in New York, it helped other refugees, particularly those from the Vienna Circle. Vienna Circle refugees also found a base in New York at the New School for Social Research, but unlike the members of the Frankfurt School, they were economically destitute. Among the refugees helped by the Frankfurt Institute in New York was Edgar Zilsel. During his traumatic exile, Zilsel maintained his views, derived from Austro-Marxism, that one should attempt to integrate theory and practice and to work for political and social change while pursuing theoretical work. In contrast, the leaders of the Frankfurt School in New York, such as Max Horkheimer, followed a new policy of restraint from direct political involvement. Despite this important difference, Zilsel wrote his first articles relating to the sociological roots of science at the Frankfurt Institute in New York. Henryk Grossmann was also affiliated with the Frankfurt School at Columbia, although he had become an increasingly marginal figure.[25]

Zilsel was familiar with the work of both Grossmann and Borkenau and seemingly dismissed the work of the latter in his article on the origins of the concept of physical laws, published in 1942. His project on the sociological origins of science, which occupied him from his arrival in New York until his suicide in 1944, had its inception certainly in his own scholarship and teaching of adults in Vienna in the 1920s but also in Boris Hessen's famous paper of 1931 and in the Borkenau-Grossmann controversy that occurred within the Frankfurt Institute in the 1930s.[26]

In addition to work on the concept of physical laws, Zilsel wrote two other influential articles in the early 1940s. The first, "Problems of Empiricism," was published in 1941, and the second, "The Sociological Roots of Science," in which he outlined his general approach, in 1942. Zilsel viewed the emergence of the new sciences of the seventeenth century as a sociological process. Crucial to that process was the transition from feudalism to early capitalism, centered in the towns and characterized by rapid technological progress and by the increased valuation of quantitative approaches. During the sixteenth century, a new group of superior artisans distinct from both university scholars and "humanist literati" emerged as an influential class. The new group comprised artist/engineers, surgeons, musical and scientific instrument makers, surveyors, and navigators. This new group brought about the new experimental methods of science when their values were accepted into academic science around 1600. The approach of those in this group included an appreciation of empirical methods and of precision measurement and quantification. Zilsel argued that these superior artisans also advocated the value of cooperation and believed that scientific progress would result. In a further article, Zilsel showed that the artisan/practitioner Robert Norman (fl. 1580s), a mariner and compass maker, exerted a decisive influence on the thought of William Gilbert, whose treatise on magnetism, *De magnete*, published in 1600, is considered one of the important texts of the new sciences. Zilsel also wrote an article in which he argued that the modern idea of progress originated in the writings of sixteenth-century skilled artisans.[27]

Zilsel positioned himself on the Hessen/Grossmann side of the debate in that he emphasized the importance of the influence of artisans, rather than the production process itself, and he placed the significant changes earlier than the seventeenth century. Yet he did not adopt the view of Hessen and Grossmann that mechanical machines such as those created by Leonardo da Vinci brought about a mechanical world view. Zilsel emphasized instead a conceptual revolution involving the positive valuation of practice, quantification, and the notions of cooperation and progress. Here he appears to have adopted Max Adler's sociological reading of Marx in which mental categories are substituted for material ones.

Leonardo Olschki and Robert Merton: Non-Marxist Proponents of Artisanal Influence

Scholars working outside of the Marxist tradition also contributed to the thesis of artisanal influence. One of the most significant, the philologist Leonardo Olschki, was the son of a prominent Italian publisher. Olschki was trained in French and Italian literature. He taught at the University of Heidelberg from 1913 to 1933, at which time he was fired after the enactment of the Nuremburg anti-Jewish laws. He then moved to Rome, where he taught until 1938, and then to the United States, where he secured a position at Johns Hopkins University and later at Berkeley. Between 1919 and 1927, he published a three-volume work on technical and practical writings in the centuries before Galileo and on the writings of Galileo. He analyzed these writings in terms of their language. As H. Floris Cohen has described it, he "examined such topics as the emancipation of the vernacular from Latin; the reasons one or another scientific author may have had for writing now in the one, then in the other language; the place of all these writings in the literary prose of their time; their stylistic, grammatical, semantic, and literary peculiarities; the gradual adaptation of a language originally quite unsuited to scientific terminology to the formal requirements of standardized scientific reporting." Olschki showed how language determined arguments. He described technical and practical writings in detail, exposing how they were based on the experience of artisans, practitioners, and travelers. This tradition of writing, he argued, gave rise to the thought and writings of figures such as Galileo.[28] Although Olschki's ideas about the significance of language were not taken up, his detailed discussions of the writings of men who engaged in technical practices—men such as the humanist and architect Leon Battista Alberti, the goldsmith Lorenzo Ghiberti (1378–1455), the mathematics teacher Luca Pacioli (1446/7–1517), the painter Piero della Francesca (c. 1415–1492), and the painter/engineer Leonardo da Vinci—played an important role in disseminating the detailed content of such writings, making them available for further utilization.

A very differently oriented non-Marxist scholar who contributed to the thesis of artisanal influence was the American sociologist Robert K. Merton, who developed his own version of the thesis in his Harvard dissertation in the 1930s. Merton was awarded a Ph.D. in sociology in 1938 from Harvard University, where he was influenced by George Sarton (1884–1956), a Belgian chemist and historian considered the founder of the discipline of the history of science. Merton also absorbed the thought of the sociologist Max Weber (1864–1920) and of the American sociologist Talcott Parsons (1902–1979). He acknowledged and followed up on the work of Boris Hessen and of Franz Borkenau. However, he rejected grand schemes of history such as those promulgated by Marxism. Rather, he was an empiricist who strove to establish middle-range theories on the basis of empirical studies. In this he was influenced by two of the founders of the discipline of sociology, Emile Durkheim (1858–1917) and Max Weber. In the first part of his 1938 *Science, Technology, and Society in Seventeenth-Century England*, Merton treats the relationship of Protestantism to the scientific revolution, basically supporting Weber's thesis of the relationship of the two movements. But in the second half, he discusses the importance of technical developments in mining, war, and transportation to the development of empirical methodologies in science.[29]

Alexandre Koyré and the Anti-Marxists

Opposing most of the scholars discussed above, many Anglo-American scholars in the 1940s and 1950s, writing from their own political and philosophical points of view, rejected the thesis of artisanal influence outright. These scholars viewed the great landmarks of the new sciences as theoretical developments, untouched by the surrounding society or the artisanal cultures within it.

Most important was the work of Alexandre Koyré (1892–1964). Koyré was born in Russia of Jewish parents, worked for a time in Paris, and subsequently moved to the United States. In 1953 he delivered a series of lectures at Johns Hopkins University that became his influential book, *From the Closed World to the Infinite Universe.* Here as in earlier writings, Koyré presented his views concerning how science had developed in the sixteenth and seventeenth centuries. He rejected the idea that new knowledge was discovered through scientific experiments. Rather, he believed that theory and ideas guided scientific research. Thus he was opposed to the positivistic view that scientific investigation should restrict itself to discovering observable phenomena, the relations between them, and the laws that directly affect them. For example, he disputed the common view that Galileo's experiments led to Galileo's conclusions. Indeed he questioned whether Galileo even did some of the experiments that he claimed he did.[30]

Koyré's work had a tremendous influence, but it also reflected the trends of Anglo-American scholarship in the history of science in the 1950s through the 1970s. Historians of science during these decades focused on great men of science, such as Copernicus, Galileo, and Newton, and their ideas, which they viewed as constituting the origins of modern science. They ignored historical facts that they did not view as "scientific," such as Newton's alchemy. Many also vociferously opposed any notion that either material culture or the artisanal world influenced the development of early modern science. These idealist views were profoundly influenced by the desire to defend the autonomy of science (and therefore the history of science), on the one hand, and by anti-Marxian views, on the other.[31]

In his 1957 paper titled "The Scholar and the Craftsman in the Scientific Revolution," the influential historian of early modern science, A. Rupert Hall (1920–2009), began by emphasizing "the great diversity of men in the forefront of [seventeenth-century] scientific achievement," including "instrument makers, opticians, apothecaries, surgeons, and other tradesmen." Yet the great period in science for these men, Hall went on to say, was the eighteenth, rather than the seventeenth, century. Hall cited Merton as the source for his statement about diversity, but he did not elaborate the possible implications. Rather, he presented a carefully reasoned argument that the scientific revolution was primarily a revolution in theory and explanation led by scholars. For Hall, neither technology nor craftsmen made a direct contribution to that revolution. It was absurd to think, he argued, that the introduction of gunpowder artillery was the cause of a revival in dynamics. After all, torsion artillery had been available since antiquity. Scientific empiricism was a philosophical artifact, the creation of learned men. It stood in the same relation to craftsmanship as the theory of evolution to pigeon fanciers. Hall did not deny that the processes of artisans constituted an important part of the natural environment. An increasingly rich technological experience offered ample problems for inquiry. Hall claimed that the great works of craft description and invention published in the sixteenth century, such as those by "Cellini, Agricola, Biringuccio, Palissy, Ercker, and Ramelli," were insignificant in terms of scientific content. Nevertheless, they provided materials and methods for the use of others "more philosophically equipped than themselves."[32]

New Approaches from the 1970s

Hall's view of the history of science as a history of ideas developed by particular individuals insulated from the culture and society surrounding them was accepted by most historians in the decades after the Second World War. Beginning in the 1980s, such views changed dramatically because of a new understanding of science that is generally referred to as constructivism. This is the view that, to quote Jan Golinski, "regards scientific knowledge primarily as a human product, made with locally situated cultural and material resources." This view of the history of knowledge about the natural world has transformed the way in which many historians of science practice their craft. Science is no longer seen as a single entity that develops in steady upward progress. Many historians now investigate what people of a given era took knowledge to be, regardless of whether it would be considered true or "scientific" in our own time. Because of its broad new range of subject matter, which includes alchemy, collecting, and magic, as well as, for example, astronomy, the history of science has become closer to other kinds of historical studies.[33]

This change involved a complicated development, in which Thomas S. Kuhn's groundbreaking book, *The Structure of Scientific Revolutions*, first published in 1962, played an important role. Kuhn's now famous argument is that there exist two phases of scientific development, normal and revolutionary. In normal science a broad consensus exists in the scientific community as knowledge about a particular focus of research grows cumulatively, and scientific research resembles puzzle solving. This normal science leads to repeated anomalies, however, which compel, as Paul Hoyningen-Huene puts it, "more or less thoroughgoing revisions of its guiding regulations." Changes occur in data about the natural world, in the empirical concepts with which the world is described, and in the knowledge implicitly contained in those concepts. Kuhn called the change from one phase of normal science to the next a paradigm shift. Kuhn's book led to decades of discussion and debate. An important core idea of his book was the view that scientific ideas were taken to be true as a result of agreement among groups of scientists, and that this consensus involved sociological circumstances rather than the relationship of a hypothesis to evidence in the real world.[34]

Kuhn's work was then taken further by David Bloor and Barry Barnes's "strong programme" at the University of Edinburgh in the 1970s, which proposed that "science should be studied like other aspects of human culture, without regard to its supposed truth or falsity." Their work led to the burgeoning field of the sociology of scientific knowledge. The Bloor and Barnes program was challenged by Bruno Latour and Michel Callon and their "actor-network" approach to the sociology of science. The complex developments of this sociological movement are outside of the purview of this book, but are detailed and debated in a large literature.[35] What is important to note here is that the sociology of knowledge influenced historians of science to develop sensitivity to the social, political, and cultural context of the topics they investigated. Particularly important for this trend was an enormously influential book published in 1985 and authored jointly by Steven Shapin and Simon Schaffer: *The Leviathan and the Air-Pump*. This book examined the seventeenth-century debate between Robert Boyle and Thomas Hobbes by trying to understand how arguments were legitimized as true and experiments taken to demonstrate valid conclusions. Here, social circumstances such as the social status of witnesses were investigated as crucial issues in the development of new ideas within what was now called experimental philosophy.[36]

These trends in the history of science brought about a reconsideration of the thesis first articulated in the 1920s and '30s that artisans and artisanal culture influenced the development of empirical methodologies within the new sciences of the sixteenth and seventeenth centuries. A group of writings has emerged dealing freshly with the subject in one way or another. Among them is Paolo Rossi's *Philosophy, Technology, and the Arts in Early Modern Europe*; Jim Bennett's work on mathematical practitioners in late medieval and early modern England; William Eamon's *Science and the Secrets of Nature*; my own *Openness, Secrecy, Authorship: Technical Arts and the Culture of Knowledge from Antiquity* to the Renaissance; and Pamela H. Smith's *The Body of the Artisan*.[37] In addition, many recent studies investigate technical practices and material culture: Paula Findlen's study of collecting; studies of alchemy and the practices of alchemy, such as those by William Newman, Lawrence Principe, and Tara Nummedal;[38] studies of natural history in early modern Europe and the new world, such as those of Harold Cook on the development of natural history in the Netherlands in the context of Dutch colonialism and markets; and the work of Brian Ogilvie, Alix Cooper, and Antonio Barrera-Osorio.[39] Finally, studies based in particular locales such as Deborah Harkness's study of the empirical cultures of London in the Elizabethan age reveal in detail flourishing practical and empirical cultures in particular locales.[40]

The influence of artisanal culture on the development of the new sciences has been an issue in the history of science since the 1920s. Although the issue was often a contentious one in the early decades of the discipline, the contenders on opposite poles of the issue actually shared significant assumptions. Most viewed the "scientific revolution" as a true revolution that occurred in the seventeenth century and that signified the origins of modern science. Whatever their views about artisanal influence, all assumed that the artifactual and the natural were separate entities, that "artisans" and "scholars" were unchanging categories, and that the individuals in question fell into one or the other category.

The following chapters explore the thesis of artisanal influence on the basis of primary sources and relevant scholarship. If artisanal culture did influence approaches to the investigation of the natural world, it is necessary to investigate anew how and in what ways that influence was exerted and to ask what conditions encouraged such an influence. As will be seen, what emerges is that several categories that were central to the issue—"art" and "nature," and "artisan" and "scholar"—themselves tended to destabilize in the fifteenth and sixteenth centuries. "Art" and "nature" came to be seen as closer together, even interchangeable. Meanwhile, certain workshop-trained artisans and certain humanist scholars increasingly shared common interests and practices, while a few, fully active in the practices of both, cannot be placed in one category or the other. The increasing proximity of art and nature, and of artisan-trained and university-trained individuals is crucially relevant, I argue, to the issue of artisanal influence.

CHAPTER 2

Art, Nature, and the Culture of Empiricism

By the early fifteenth century fundamental cultural changes were under way that gave fabricated objects and the people who made them greater social and cultural significance. One result was that gradually the worlds of artisanal practice and the worlds of learning moved closer together. This growing proximity eventually allowed the values and practices of artisans to influence the culture of learning, including investigations of the natural world. Values intrinsic to artisanal work—entailing empirical approaches, an appreciation for individual experience, hands-on practice, instrumentation, observation, and measurement—became more readily available for use as methodologies for the investigation of nature. As part of this development, "art," in the sense of the productive arts or crafts, and "nature" came to be perceived as closer together, at times interchangeable.

"Art" in the premodern world was not art in the modern sense, referring to fine arts such as painting and sculpture that today are perceived as different from practical skills such as carpentry and weaving. In the premodern centuries, "art"—or *ars* as it would be called in Latin; or *technē*, to use the equivalent Greek word; or *artes mechanicae*, the mechanical arts, as they were often called after the ninth century—referred to the crafts, handwork, and practices used, for example, in the construction of buildings and in navigation. Painting and sculpture were included among such activities in the medieval period and were not distinguished from them. Gradually, in the fifteenth and sixteenth centuries, painting and sculpture rose in prestige. Often they came to be considered mathematical arts due to the invention of one-point artist's perspective in the early fifteenth century and to the growing interest in classical proportions in sculpture. By the end of the sixteenth century, as Paul Kristeller pointed out in a famous article, painting and sculpture had begun to transform into the modern system of the arts, or "fine arts."[1]

Art and Nature: Changing Concepts in History

As "art" changed, so also did "nature." The changing nature of nature and its changing relationships with art are the focus of new scholarly attention since the 1990s. In an insightful essay, Lorraine Daston outlines the shifting relationships among the supernatural, the preternatural, the artificial, and the unnatural as "forms of the non-natural that bounded and defined the natural."[2] Nature was not a single reified entity referring to a stable "reality" over the centuries. Rather, it was a cultural construct that shifted in meaning from one milieu to the next and from one era to another. In a collection of essays edited by Bernadette Bensaude-Vincent and William R. Newman, scholars discuss particular contexts that range from antiquity to the modern world and that reveal varying meanings and relationships of art and nature. In a masterly essay, Heinrich von Staden discusses the changing meanings and relationships of *physis* (nature) and *technē* (art) in ancient medical traditions from the Hippocratic tradition in the fifth and fourth centuries BCE to the third-century-BCE vivisectionist Erasistratus and beyond. Mark Schiefsky provides a study of the issue within ancient mechanics. He shows that the traditional view that Aristotelians believed that mechanics was contrary to nature and thus could not be studied as a way of understanding nature is simply incorrect. He provides a more complicated analysis, showing that often Aristotelian mechanical writings portrayed nature and art as closely analogous.[3]

Such recent studies suggest that the traditional generalization is no longer sufficient—that Aristotle believed that art and nature were quite separate and that art was inferior to nature, rendering it incapable of being compared with or used to understand nature. Aristotle's writings do provide key texts, partly because they were so widely known in the medieval era. In one well-known passage, he sharply distinguishes natural things from those made by human hands—between nature and art, or *physis* and *technē*.

> Of things that exist, some exist by nature, some from other causes. By nature the animals and their parts exist, and the plants and the simple bodies (earth, air, fire, and water). . . .
>
> All the things mentioned plainly differ from things which are *not* constituted by nature. For each of them [i.e., each of the things of nature] has within itself a principle of motion and of stationariness (in respect of place, or of growth and decrease, or by way of alteration). On the other hand, a bed and a coat and anything else of that sort, *qua* receiving these designations—i.e., in so far as they are products of art—have no innate impulse to change.[4]

Thus, things of nature can move and change in and of themselves, whereas products of art cannot. In addition, Aristotle posits two kinds of art—one that can improve upon nature and another that "imitates" nature: "generally art in some cases completes what nature cannot bring to a finish, and in others [art] imitates nature."[5]

This bifurcated view of fabricated things and their relation to natural things—imitative and therefore inferior, on the one hand, or able to actually improve upon nature's own processes, on the other—fueled an intense debate concerning alchemy in the medieval period. As Newman points out, the Persian thinker Avicenna (Ibn Sīnā, 980–1037) argued that the alchemical transmutation of metals was not possible because the artifice and art used to attempt such transmutation were inferior to nature and thus not able to improve upon it. His anti-alchemical stance was reiterated and elaborated by other Arabic writers. This led to counterarguments by Roger Bacon (1214/1220–1292) and Thomas Aquinas (1225–1274) in the late thirteenth century and to a full-scale defense of alchemy in alchemical texts such as the *Theorica et practica*, possibly by Paul of Taranto, a thirteenth-century Franciscan. The defense of alchemy entailed the argument that human art in the form of alchemy could create better and more plentiful things than could nature itself—art could correct and improve upon nature.[6]

Scholarship in the history of alchemy, especially by Newman and Lawrence Principe, investigates the relationship of alchemy to the empirical practices of seventeenth-century figures such as Robert Boyle (1627–1691). The view that alchemical art could have transformative effects on the natural world became part of an influential tradition that advocated and practiced empirical methodologies and trusted the efficacy of art (i.e., artisanal skill) for both investigating and changing the natural world. Newman and Principe have demonstrated the importance of the alchemical tradition to the new sciences of the seventeenth century, especially to such canonical figures as Boyle.[7]

Focusing on the German states, Tara Nummedal shows the extensive overlap of alchemy and alchemical laboratory operations with practical crafts and productive arts such as mining and metallurgy, including minting, assaying, and goldsmithing. She also investigates the work and lives of particular individual alchemists in the German states, their patronage, laboratories, and connections to the economic ambitions of princes.[8] Earlier, the pioneering work of scholars such as Betty Jo Teeter Dobbs helped restore the extensive alchemical writings of Isaac Newton to the legitimate corpus of his serious work.[9] Alchemy has thus been reevaluated. No longer an "occult" subject that has nothing to do with "science," it is now understood as playing an important role in the development of the empirical methodologies of the seventeenth century.[10]

Experience and Experiment

Beyond the discipline of alchemy, to what extent did medieval people engage in observation and experiment as methods of investigating the natural world? First, it must be said that experiment and empiricism were never entirely absent from medieval culture. Ancient empirical traditions continued in medicine through Hippocrates (ca. 460–ca. 370 BCE) and Galen (ca. 131–ca. 201 CE) and beyond; in astronomy through the work of Ptolemy (ca. 100–ca. 170 CE) and then Arabic observational traditions; and through optics and other fields in the work of such figures as Robert Grosseteste (ca. 1168–1253), Ibn al-Haytham, known as Alhacen (965–1040), and Roger Bacon. The extent and meaning of empiricism and experimentalism within such traditions is still an issue.[11] What was understood by experience and experiment in the medieval centuries? How extensive and thoroughgoing were they as practices?

Such questions have provoked disagreement among historians of science. Newman contends that Aristotle's separation of art and nature has led to what he calls the "noninterventionist fallacy," that is, the view among several prominent historians of science that Aristotle and his followers were "fundamentally nonexperimental or even actively opposed to experiment, because experimentation involved intervention in natural processes."[12] Newman points to three historians of science who he believes have erred in this way—Peter Dear in his *Discipline and Experience* and Lorraine Daston and Katharine Park in their *Wonders and the Order of Nature*.[13]

Newman points out that Aristotle himself did what could be called experiments, for example, in his studies of chick embryos and in his study of rainbows. Nevertheless, I would not go so far as to call Aristotle filling "the role of an experimental scientist" as Newman does, in part because the phrase "experimental scientist" is anachronistic.[14] Beyond this quibble, I would say that Aristotle's experimental and observational approach, especially in his zoological work, is not really a point of contention among historians. Most historians of science agree that Aristotle did do experimental or empirical work, especially in his study of animals. Nor have I seen anywhere that Dear, Daston, or Park denies that Aristotle himself did "experiments," although their meaning and how they fit into his philosophical system as a whole are rightly a focus of extended discussion, as is their influence.[15] But the question is this: To what extent did medieval Aristotelians use experimental procedures in fundamental ways as Aristotle did in some of his biological and meteorological works? It is clear that the procedures of most of the Aristotelians working within the scholastic frameworks of medieval universities usually were based on the commentary and teaching of authoritative texts and grounded in the analysis of causes, and that their work was not based on experiment or observation in fundamental ways.

Dear argues in *Discipline and Experience* not so much about results of the perceived separation of art and nature as about the difference between how Aristotelians viewed experience and how experimental philosophers of the seventeenth century viewed experiment. Dear suggests that ordinary experience, that is, experience that always occurs in given circumstances (if you drop a brick from a height it will always fall to the ground), was for Aristotelians the basis of legitimate knowledge about the natural world. This was in contrast to individual, constructed experiments that draw conclusions from highly manipulated and idiosyncratic operations conducted by specialists or experts, using, say, a complex apparatus such as an air-pump. Robert Boyle had just such a complicated machine constructed (by his assistant Robert Hooke, 1635–1703), which he used for experiments on air. It was an apparatus that could be constructed and operated only by experts. Only they could draw legitimate conclusions from the experiments. Dear's central argument is that Aristotelians in late medieval Europe gave more credence to the first kind of experience, that is, ordinary experience that everyone could agree on, than to the second, experiment, which set up special circumstances and could be evaluated only by individual experts.[16]

Park and Daston, in *Wonders and the Order of Nature*, instead trace a complex and gradual collapse of the distinction between art and nature, a distinction that had in general prevailed before the sixteenth century. The proximity of the two categories of art and nature can be seen—to use an example discussed by Pamela H. Smith in *The Body of the Artisan*— in the platters made by the sixteenth-century potter Bernard Palissy in which he embedded lizards and other once-living things into his ceramic ware;[17] in the human portraits painted by Giuseppe Arcimboldo, such as the allegory of the element fire in which the head is constructed of metal

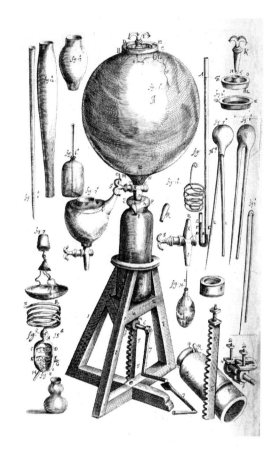

2.1. Robert Boyle's air pump constructed by Robert Hooke. From Robert Boyle, *New Experiments, Physico-Mechanical* (London: Miles Flesher, 1682). Reproduced by permission of The Huntington Library, San Marino, California.

2.2. Bernard Palissy or follower, earthenware dish with decorations in relief of reptiles, plants, and shells. © Victoria and Albert Museum, London.

weapon parts and natural materials; and in the intermingling of natural and fabricated objects in numerous natural history collections that were created in the sixteenth and seventeenth centuries. This new proximity of what had been perceived as two separate categories was a widespread cultural phenomenon from the fifteenth century to the early seventeenth century.[18]

In these centuries, interest in objects and things was fueled by the rise of conspicuous consumption—an increasingly important social and political necessity for Europe's elite classes. Visible consumption required the purchase of fine clothing, furs, and jewelry made of precious metal and stones; the construction of sumptuous palaces and gardens; and the acquisition of sculpture, painting, ceramics, and furniture to furnish and ornament them. Worldly goods, as Lisa Jardine aptly described it, proliferated as they also gained in cultural value.[19]

Along with this proliferation of objects came the expansion of writing about the arts, from painting, sculpture, and architecture to guns and fortification to pottery and silk-making. Within this literature, which was created by artisan/practitioners and learned people alike, values such as hands-on skill, personal observation, a belief in the efficacy of instrumentation, the practice of using instruments and devices for measurement and experiment, and the value of individual experience— all implicitly or explicitly gained validation in the wider culture. These writings conveyed values that make up what has been called the "maker's knowledge tradition"—a tradition that informed the new sciences of the late sixteenth and seventeenth centuries and that was being articulated in one form or another long before Francis Bacon's writings in the 1620s.[20]

A closer look at several specific examples illustrates some of the ways in which the relationships of art and nature were framed and understood in the late fifteenth and early sixteenth centuries. The examples differ one from the other, but they each assume important relationships between nature and art, and in some instances, the conviction that art or the mechanical arts (in the form of machines) can be used to understand aspects of the natural world, namely power and motion.

2.3. Giuseppe Arcimboldo, fire allegory. 1566. Portrait head depicting a construction with metal weapon parts and natural materials. Kunsthistorisches Museum, Vienna, Austria. Photo credit: Erich Lessing/Art Resource, NY.

Objects of Art/Objects of Nature in a Late Fifteenth-Century Romance

The romance called the *Hypnerotomachia Poliphili,* written in the 1470s and published in 1499, is one of the most famous books of Renaissance Europe. Published by the pioneering humanist printer Aldus Manutius (1449/50–1515), its striking woodcut illustrations and typography have influenced typographers and publishers for centuries. The author is usually assumed to be a Dominican monk, Francesco Colonna (1433/1434–1527), who lived at the monastery of SS. Giovanni e Paolo in Venice. Colonna's romance encompasses a dream within a dream in which one Poliphilo walks through a varied terrain filled with pyramids, ruins, forests, meadows, streams, palaces, and gardens as he searches for his lost love, Polia. It is written in a bizarre tangled Italian that uses many Italianized Latin words taken from ancient texts such as Ovid's *Metamorphoses*, and most importantly, Apuleius's *Golden Ass*.[21]

The *Hypnerotomachia* is interesting because of the author's persistent habit of describing objects of nature and objects of art in profuse detail, including architectural fragments, and because of his habit of interchanging constructed artifacts and natural things. For example, Poliphilo comes to an enclosed valley where he encounters an enormous half-ruined pyramid carved out of the surrounding mountains and topped by an obelisk that is dedicated to the sun. The author provides a detailed description of the monument, including structures, ornamental details, fragments of sculpture and architecture, inscriptions, and the names of many varieties of herbs and grasses growing among the ruins. Poliphilo hears frightening groans and discovers that they are emitted by the colossal statue of a man. The groans are caused by wind blowing through the open mouth of the colossus. Using the hairs of the chest and beard of the statue, Poliphilo pulls himself into the open mouth and then climbs through the viscera. Each part of the body, "intestines, nerves and bone, veins, muscles and flesh," is present and each is labeled in Chaldean, Greek, and Latin. The inscribed body parts describe what sickness is generated in each part, the cause, the required care, and the remedy.[22]

Poliphilo emerges from this polyglot body and looks at other ruins, such as the colossal statue of a horse that "seemed almost to tremble in its flesh, . . . more alive than fabricated." He enters the pyramid through the great portal, encounters a dragon and other amazing things, and emerges from the bowels of the great structure into a lovely meadow. He describes the meadow by providing an encyclopedic enumeration of the specific varieties of plants and trees. As he continues on his journey he describes an ancient bridge over a stream with overgrown banks populated with many varieties of birds, and then a field filled with creatures, flowers, and fruit trees, each of which he specifies by name.[23]

He meets the nymphs of the five senses and describes their clothing minutely. He finally arrives at the palace of the Queen Eleuterilyda (Free Thought), and describes the approach to the palace where the nymphs take him first to the baths. As he draws near the palace and enters, he describes the courtyard, the exterior, the ornamentation of the entrance, the ornate tapestries of the successive rooms, the decorations, the throne of the queen, her clothing and jewelry, and each item of the lavish bejeweled service used at the sumptuous banquet. After the fabulous banquet, which is described in detail, the nymphs lead Poliphilo into a lower courtyard where they walk through an orange grove and a series of extraordinary gardens, one made entirely of glass and gold, a second that is an aquatic labyrinth, and a third made of silk and decorated with pearls and vines of gold. The glass garden contains garden pots that hold, instead of living plants, artificial ones. The author explains that in place of greenery "every plant was of very pure glass, excellently [made] beyond what one could imagine or believe, topiary box trees with the roots and stems of gold." Here is just one of many examples in which natural things and things fabricated by humans are interchanged to the delight and stupefaction of the wandering Poliphilo.[24]

The *Hypnerotomachia Poliphili* is a literary work by an author who is enthralled by architecture, antiquity, fabricated objects, and myriad natural species. It contains numerous detailed lists and encyclopedic

2.4. Francesco Colonna, *Hypnertomachia Poliphili* (Venice: Aldus Manutius, 1599), fol. A3v. Poliphilo is lost in a dark wood. Reproduced by permission of The Huntington Library, San Marino, California.

descriptions of both natural and artifactual objects. Frequently, Poliphilo expresses astonishment and amazement at both kinds of objects. The examples mentioned here—the pyramid that is actually a carved-out mountain, the colossal statue of a man that groans, the fabricated horse that trembles as if alive, and the gardens made of beautifully fashioned glass, gold, silk, and pearls—transgress the boundaries between the artifactual and the natural. The *Hypnerotomachia Poliphili* exhibits interchangeability between natural and constructed things as it also displays a kind of descriptive exuberance in which plants, birds, and trees are described in great detail, as are architectural ruins, structures, inscriptions, and crafted objects of all kinds.

Francesco di Giorgio and Humanist Engineering

A second example is a very different one, taken from the treatises of Francesco di Giorgio (1439–1501), one of the most prominent and successful architect/engineers of the late fifteenth century. Francesco trained as a painter in a workshop in Siena. Eventually he became a client of princely patrons such as Federico Montefeltre in Urbino—patrons who headed courts that were influenced by humanism.[25]

Francesco's ability to garner such patronage was due to his technical abilities but undoubtedly also to his literary efforts. He was deeply interested in Vitruvius's *De architectura*, the only fully extant architectural treatise from antiquity (which will be discussed in the next chapter). Francesco's writings include two major treatises on architecture, military engineering, and machines.[26] He wrote the first, *Trattato I*, while working in Urbino between 1477 and 1480, and the second, *Trattato II*, later, in the late 1480s or perhaps the 1490s.[27]

A comparison of the two treatises in their treatment of machines shows how Francesco moved from the culture of practice to the culture of learning. *Trattato I* is a detailed treatment that reflects the concerns of a practicing engineer. *Trattato II* sets out the topics according to general principles and follows some of the practices of humanist authorship.

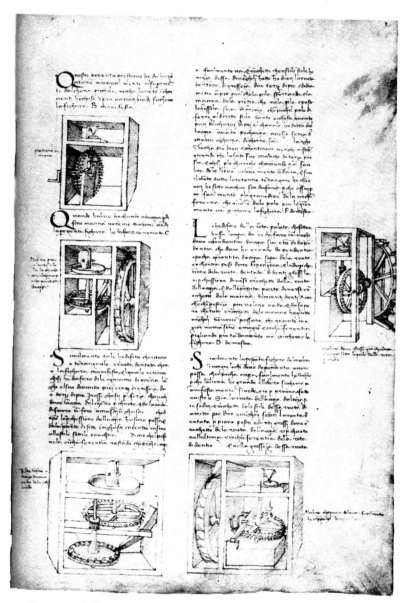

2.5. Francesco di Giorgio, *Trattato I*. Mills. Turin, Biblioteca Reale, Codice Saluzziano 148, fol. 34r. By kind permission of the Ministero per i Beni e le Attività Culturali, Biblioteca Reale, Torino.

This development can be illustrated by comparing his treatment of mills in *Trattato I* and *Trattato II.*[28]

Trattato I contains several sections that focus specifically on machines —military machines, cranes and lifting devices, and mills of various kinds. The section on mills takes up seven folios, or fourteen pages. The text, written in carefully blocked-out columns, two columns per sheet, was clearly written by a professional scribe. But scholars agree that the detailed technical descriptions make it certain that Francesco was closely involved in the creation of the manuscript. Interspersed with the columns are box-shaped drawings of mills with their wheels and gears carefully depicted. There are a total of fifty-eight such drawings of mills, which, taken as a whole, explain many variations of this ubiquitous and important machine. There are water-powered mills, including those with horizontal, overshot, and undershot wheels, each in a variety of configurations. There are dry mills powered by animals, mills turned by cranks, and windmills.

In the text placed under the drawings, Francesco carefully explains what type of mill it is and then goes on to describe the wheels and gearing, giving what he considers to be the appropriate measurements, and other details such as the best number of rods on lantern gears. In the case of waterwheels, he specifies which kind of wheel is appropriate for specific situations given the availability of water. If the water supply is somewhat

2.6. Francesco di Giorgio, *Trattato I.* Mills. Turin, Biblioteca Reale, Codice Saluzziano 148, fol. 34r. Photoshop enlargement of mill E (lower right) of figure 2.5.

sporadic, overshot wheels are better, whereas undershot wheels work well for situations where a continuous, strong flow of water is available.[29]

Take, for example, mill E, the one on the bottom right of figure 2.5, enlarged in figure 2.6. The mill is powered by an overshot waterwheel. The water spills from a funnel on top of the wheel, turning the wheel. On the shaft of the wheel is a lantern gear that rolls over the vertical teeth placed around the top circumference of the horizontal crown gear wheel. In the back, another lantern gear is attached to the shaft that turns the millstone. This lantern gear is rolling on the horizontal teeth of the same wheel. Francesco has also drawn the mill's wooden frame and the grain funnel into which the grain to be ground into flour is to be poured, and the millstones. The latter two components are drawn in disproportionately small sizes.[30]

In *Trattato I* Francesco provides a detailed written account, illustrated by numerous drawings that display his engineering knowledge. Mills were essential elements of many building projects from castles to forts, and they were also essential to any town or city. Francesco provides numerous examples of variations. Knowledge of such variations was essential to any practicing architect/engineer of his time. Presented with variations of geography, power supply, and potential use, the master must devise or decide upon a mill with particular characteristics for specific sites. Francesco's careful detailing of numerous variations points to his close identification with the concerns of practicing engineers. *Trattato I* reflects the world of practical engineering and presents it to elite patrons.

Francesco's later treatise, *Trattato II,* is organized differently. It contains fewer chapters, and erudite introductions have been added. Also added are numerous citations from ancient texts, such as Pliny's *Natural History* and the *De architectura* of Vitruvius. In addition, Francesco has moved from detailing many particular mills to more general accounts of far fewer mills.[31] This approach is more suitable for a readership of patrons and other nonpractitioners. They might be more interested in how a mill works in general than in the myriad variations appropriate to different mill uses and different locations.

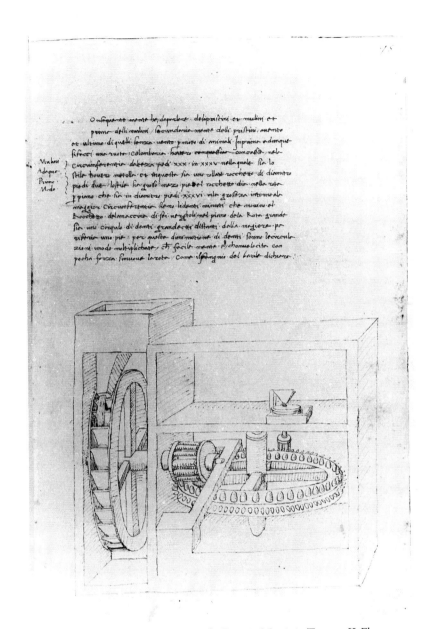

O infiquante mente he, dopochur deluposici333 et mulin et
primi delli mulin secundariamente delli pristini mente
et ultime di quelli senza uento pinori di minali Inprima adunque
fitroci una ruota colombaria hauera compustur concordsi nel
circumferentia daltezza piedi xxx in xxxv nellaquale sia lo
stile houera metolla et inquesta sia una villano rocchetto di diametro
piedi due lostilo sia quesso mezzo pie del rocchetto dia mella rota
pimo che sia in diametro piedi xxxvi nela grossezza intorno al
maggiore Circumferentia siano hidenti minuti che muoue et
Rocchetto delmacouna di sei ruzoholi nel pimo dela Rota grande
sia uno circulo di denti grandetti distanti dalla maggiore pe
rifterra una pie per quessa diminuzione di denti sonno lecirculi
zioni modo multiplicate et facile mente si chommelocita con
pocha frezza simuoue larota Come ilspingnie del barile dichiara

2.7. Overshot mill. From Francesco di Giorgio Martini, *Trattato II*. Florence, BNC, Codice Magliabechiano II.I.141, fol. 95r. By kind permission of the Ministero per i Beni e le Attività Culturali della Repubblica Italiana/Biblioteca Nazionale Centrale di Firenze.

In this later treatise, the material on machines that was scattered among various sections of *Trattato I* has received a new treatment in one section only. This section concerns lifting machines of various kinds and mills. In all only eighteen machines are displayed in drawings and described, a radical reduction from the numbers of machines found in *Trattato I,* where there are fifty-eight drawings of mills alone. Offering a rationale for this new approach, Francesco explains: "Thus again I will provide a drawing [*la figura*] of some mills, so that through those, other similar ones may be able to be discovered by readers." He thereby suggests that readers will be able to read about one kind of mill, seemingly understand its principles, and then discover other types.[32]

Francesco discusses only one type of overshot waterwheel mill, which he draws in a large size on a single page underneath the text describing it (figure 2.7). This machine is similar to the overshot wheel discussed above. His description of the mill in *Trattato II* is similar, but not the same as that for the analogous machine in *Trattato I.* In *Trattato II* he gives the dimensions of various parts such as the waterwheel and the crown wheel. He explains that the wheel turns easily because the vertical teeth on the top rim of the crown gear are larger than the horizontal teeth on the lateral rim, and there are fewer of them. The smaller teeth on the lateral rim of the crown wheel turn the smaller lantern wheel attached to the axle that turns the millstone. The drawing clearly shows the difference in size and number of the two rows of teeth on the crown gear wheel. It also shows the different sizes of the two lantern gears, the one on the axle of the waterwheel being larger than that on the millstone axle.[33]

Francesco explains that these differences in the size of the teeth in the crown wheel and the size of the lantern gears are what make the machine turn easily.[34] It is notable that in his earlier drawing of the analogous machine such differences are not apparent in the drawing, nor does the explanation concerning ease of motion appear in the earlier text. This is an example, then, of using one machine to explain a more general principle (that of the gear differential), instead of enumerating varieties of mills suitable for diverse situations as he did in *Trattato I.*

The changes that are apparent between *Trattato I* and *Trattato II* indicate Francesco's efforts to write a more learned treatise, efforts that were influenced by humanist practices of authorship. The differences between Francesco's two treatises can be explained by his increased interest in and ability to address humanist learned culture. This development has significance not only in terms of Francesco di Giorgio's own career, but because it is part of a larger development in which practical and technical knowledge was becoming integrated into written and learned traditions. Put in another way, Francesco at first used his art—that is, his knowledge of engineering and mills—for the sake of that art, in order to elaborate in detail the various kinds of mills that should be used in diverse situations and sites. Subsequently in *Trattato II*, he deployed his knowledge for very different purposes—first to provide a general understanding of machines and mills to the unskilled, and second, to contribute to natural knowledge as it pertained to the power and motion of machines.

Leonardo da Vinci: Elements of Machines and the Study of Motions

Leonardo da Vinci, who was a contemporary of Francesco di Giorgio, was trained in the workshop of Andrea del Verrocchio (ca. 1435–1488) in Florence both as a painter and sculptor. At this time in the late 1460s, the towering double-shelled cupola of the Florentine cathedral Santa Maria del Fiore, created by Filippo Brunelleschi, was completed except for the heavy sphere of gilded copper that was to go on top of the lantern. The placement of the sphere was given to Verrocchio's workshop, and the work was completed between 1468 and 1472. Brunelleschi's innovative lifting machinery and some of the scaffolding were still in place and available for examination, and Leonardo studied them thoroughly.[35]

Leonardo's fame as a painter and engineer was matched by his "literary" production. He filled an estimated twelve thousand sheets in notebooks in his lifetime, about half of which are extant. The two notebooks referred to as the Madrid Codices had been miscatalogued in the Biblioteca Nacional in Madrid for centuries and were rediscovered only in the winter of 1964/65. The first of the notebooks, called *Madrid I*, consists of a coherent treatise about the elements of machines and mechanics. It is filled with beautiful drawings of parts of machines and other devices, as well as notes that discuss their motions and workings.[36]

Here I examine in greater detail just one of these pages, folio 15v, with the heading "Of pinions and wheels." On the page Leonardo has drawn numerous variations on the topic of pinions and wheels (by which he means gears). Both the drawings and the textual explication allow analysis of individual mechanisms and motions as well as comparisons among them. Leonardo begins with a statement that concerns the construction of the devices. "If the pinion must move the wheel, the spacings of the pinion's teeth must be wider than those of the wheel's teeth. And if the wheel turns the pinion, the opposite should be made. However, if both were well constructed, having equal teeth and spacing would be satisfactory." The text accompanying individual figures concerns the relationship of the structure of the gears to the direction of the motion.[37]

On the top right (see figure 2.8), under the spur gears driven by a weight, Leonardo writes that if you turn one of the wheels in one direction, the other will turn in the opposite direction. The drawing below (second figure in right column) depicts three interlocking spur gears. Leonardo advises that "if you wish to have a wheel that turns in the same direction as does the movement of the first wheel, this wheel, by necessity, must have a motion of the third degree," that is, it must have an additional gear in the center, as he has shown.[38] Other depictions show other variations, for instance (third figure in right column), a weight-driven rack turning a spur gear. At the bottom right he shows three encased wheels and writes: "Regardless of the direction you turn wheel d, a and b shall turn the same way. And wheel c will turn contrariwise. The same will happen when turning a and b. But if you turn wheel c [the center wheel], every other wheel will turn in the opposite direction."[39]

Of the device that is fourth down in the left column, he says, "But if one of the wheels turns the inside of the other wheel with its outside, both wheels will then turn in the same direction regardless of which wheel causes the motion." He shows (and sometimes explains the

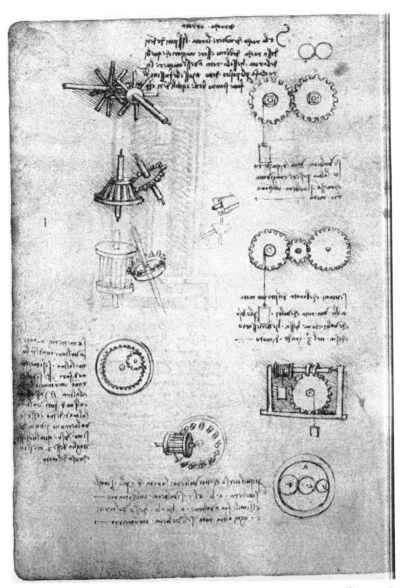

2.8. Leonardo da Vinci, "Of Pinions and Wheels." Madrid Codex I, fol. 15v. ©
Biblioteca Nacional de España, Madrid.

motions of) other devices as well, including crown and lantern gears, and other kinds of gearing.[40]

Leonardo thus devises a number of variations, and he studies the differences that these variations make in the direction of motion. He is interested in both structural variations of different kinds of pinions and gears and in the local motions that result from these variations. These issues have practical manifestations for real machinery, but Leonardo is also interested in the study of motion itself in all its small, local manifestations. These drawings, of which there can be little doubt that they were drawings of actual working gears that he had constructed, can be thought of as observational tools.

Leonardo is here observing the motions of fabricated objects. He has had gears of various forms constructed and has placed them in various positions. Artisanal craftwork and the study of nature, that is, the study of motion, go hand in hand. This is not to say that Leonardo's studies in *Madrid I* led directly to a new mechanics. In fact, Galileo's achievements of slightly more than a century later were possible because of Galileo's ability to abstract motion and think of it mathematically.[41] In contrast, Leonardo's studies were highly specific and indeed the opposite of abstract. They were tied to the observation of the behavior of particular gears and gear arrangements. Leonardo pursued a kind of observational mechanics that was thoroughly empirical, but would not lead to the modern science of mechanics.

Serlio on Buildings and Vesalius on Bones

The growing proximity of the artificial and the natural was facilitated by the increasing interaction of artisanal and learned cultures in the cities and courts of Europe. An example of this can be seen in the activities of two men, Sebastiano Serlio (1475–1554), a man trained as a painter who studied buildings and wrote architectural books,[42] and Andreas Vesalius (1514–1564) a university-trained physician who taught at the University of Padua.[43] Serlio wrote a series of treatises on architecture, the first published in 1537. His writings may have influenced Vesalius in his approach to the *De humani corporis fabrica libri septem* (*On the Fabric of the Human Body*), his renowned treatise on human anatomy, published in 1543. In their respective topics of architecture and anatomy, these writings pioneered new relationships between the use of visual images and their integration with explanatory texts. Although specific contact between the two men is undocumented, Vesalius was undoubtedly familiar with Serlio's treatise on architecture. Serlio was trained as a painter in Bologna, spent some years in Rome, and moved to the Veneto in the 1530s, becoming associated with the circle of the renowned Venetian painter Titian (ca. 1488–1576). Titian's circle of friends included painters as well as literary figures such as Pietro Aretino (1492–1556), a satirist, poet, and playwright. Vesalius also lived in Padua and Venice in the late 1530s, and he taught anatomy at the University of Padua, where he also wrote his treatise on anatomy that came out of his public dissections of human corpses at the anatomy theater of the university. My comparison of the two works takes up a suggestion by Vaughan Hart and Peter Hicks that Serlio's treatise bears notable similarities to Vesalius's *Fabrica*.[44]

The identity of the illustrators of the *Fabrica* is unknown, but it is certain that they were closely supervised by Vesalius himself—the extremely detailed anatomical descriptions and numerous cross-references between images and text make this indisputable. The Flemish artist Jan Steven van Calcar (?1499–?1546), who created the illustrations for Vesalius's earlier work, the *Tabula anatomicae sex* (1538), also executed at least some woodblocks for the *Fabrica*. Others also seemed to have worked on Vesalius's great masterwork, some of them apparently more skilled than Calcar.[45]

More important than the specific authorship of the images are the substantial ties that bound the *Fabrica*—and thus the developing discipline of medical anatomy—to the worlds of painting and sculpture. The remarkable images of the *Fabrica* not only were created by trained painters but engaged the culture of painting in substantive ways. For example, the illustrations of a series of visceral figures in book 5 are based on the famous Belvedere Torso, an antique sculptural fragment

2.9. Andreas Vesalius, *De humani corporis fabrica libri septem* (1543), 372. Torso. Courtesy Rosenwald Collection, Library of Congress, Washington, D.C.

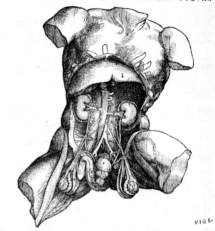

VIGESIMASECVNDA QVINTI LIBRI FIGVRA

2.10. The Belvedere Torso, Apollonios of Athens, first century BCE. Museo Pio Clementino, Vatican Museums, Vatican State. Photo credit: Scala/Art Resource, N.Y.

first reported in the collection of Cardinal Prospero Colonna in the early 1540s; thus did Vesalius counter the messy and revolting reality of abdominal dissection with the frame of a sculpted, ideal human form. Similarly, Vesalius aimed, it has been argued, at "Polycletian bodies," those that conformed to the standards of the ancient Greek sculptor Polycletus (fl. ca. 450 BCE), who created a statue displaying a canon of ideal human proportions.[46]

Serlio's first publication, which appeared six years before the *Fabrica* in 1537, treats the five architectural styles or orders. The architect meant his didactic handbook for both designers of buildings and for others who were interested in Vitruvius and in classical building styles. It is an illustrated treatise with pictures on virtually every page depicting various parts of buildings designed in a variety of ways. One example illustrates and also describes three different gates. The first, called rustic work, is suitable for a country villa; the second, suitable to the Tuscan style, was seen in Trajan's Forum for a long time, although it is now in ruins; and the third is a door with "a segmental arch, which is the sixth part of a circle, a work of great strength." Serlio in this way depicts various kinds of architectural styles pertinent to particular elements of the building, many if not all from observation. He shows them to the architect, who can then pick and choose various elements for his own designs. As Serlio says about his drawing of the gate from Trajan's Forum: "The two niches on either side are out of place, but I have put them here in order to demonstrate the different types of niches which would suit such work, so that the judicious architect can make use of them and put them in the right place."[47]

Serlio often seems to work with Vitruvius's *De architectura* in one hand and ancient architectural fragments and ruins in the other. For him, the discipline of architecture involves the detailed study of the ancient text, careful observation of ancient structures and ruins, and depiction of those observations in the form of carefully rendered illustrations. He writes, "Because I find great discrepancy between the buildings in Rome and other places in Italy and the writings of Vitruvius, I wanted to show some elements which, to the great pleasure of architects, can still be seen

E *t perche io trouo gran differentia da le cose di Roma & di altri luochi de Italia, a i scritti di Vitruuio. Ho uoluto
dimostrarne alcune parte de lequali si ueggono anchora in opera con gran satisfation de gli Architetti, & benche elle
siano di picciola forma, & senza numeri & senza misure, non dimeno sono proportionate alle grandi & con grā di,
ligētia da grāde a picciole traportate, il capitello, R, fu trouato fuor di Roma ad uno pōte sopra il fiume detto Teue
rone, il capitello, V, e in Verona sopra un arco triōphale, il capitello, T, è ad un tēpio dorico al carcer Tulliano i Ro
ma, il capitello, P, fu trouato a Pesaro cō molte altre cose antique degne di lode. La pittura del quale, anchor ch'el
la sia grande, non dimeno è molto grata a i riguardanti, il basamēto la base il capitello, A, sono al foro Boario in Ro
ma, la cornice il capitello & la ipoista di un arco, B, sono al i beatro di Marcello, la cornice fregio & architraue, A,
sono al foro Boario in Roma, lequai tutte cose ho uoluto dimostrare, accio che lo Architetto possi fare elettion di quel
che piu gli ograda in ȝsto ordine dorico hora seguitarò in tal spetie alcune particular misure necessarie a l'Architetto.

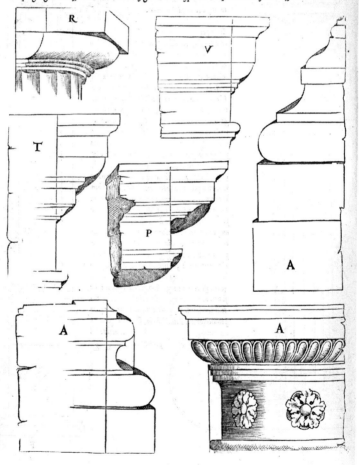

2.11. Sebastiano Serlio, *Regole generali di architectura sopra le cinque maniere
de gliedifici . . .* (Venice: F. Marcolini da Forli, 1537), fol. XXIv (mislabeled
IIIv). Elements of the Doric order. Courtesy Rosenwald Collection, Library
of Congress, Washington, D.C.

molles esse: in grandioris uerò ita subinde indurari, ut fragilis ac friagilis ossis naturam refe-
rant.quod potissimum laryngis cartilaginibus,& illis accidit quas superiores costæ educunt.
Hæ enim temporis successu, idẽ maximè in brutis, osseæ fiunt, exterius duntaxat cartilagine
ueluti membrana quapiam succinctæ, quæ ab ossea illa cartilaginis substantia per elixationem
leui negotio abscedit & diuellitur.

NOMINA QVIBVS OSSIVM PAR-
tes sedesẽ indicantur. Caput III.

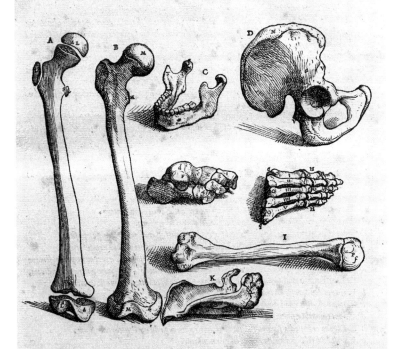

PRAESENTIS TABVLAE EIVSDEMQVE
characterum index.

*Proposita modò tabula aliquot ossi in hoc tantù delineantur, ut ossium partes & sedes, qua-
rum nomina hoc Capite persequar, in nonnullis saltem ossibus appositè exprimant. Vndestin
huius Capitis contextu alicuius ossis hic non delineati mentio incidet,id ex figuris proprij sut
Capitis, aut ex integris quæ primi libri calci adhibebuntur, opportunè petes. Quanquam non
opus est,in cuiusque partis sedisẽ descriptione,omnia ossa, quæ exempli loco adferam,contem
plari,quum hic abunde erit in uno osse quod traditur perpendere, & quum alibi id nomen occur-
ret,memorem esse. Quæ uerò ossa hic delineantur,characterum index docebit.*
A *Dextrum femur, cuius appendices sua sede dimouimus, ac ueluti quodam interstitio à reliquo
osse disiunximus.*
B *Dextrum femur, cui appendices suis sedibus adhuc adnatæ hærent.*
C *Maxilla inferior,una cum humiliori dentium serie.*

2.12. Andreas Vesalius, *De humani corporis fabrica libri septem* (1543), 5.
Collection of bones. Courtesy Rosenwald Collection, Library of Congress,
Washington, D.C.

on buildings." He provides pictorial representations of various elements of different buildings that he has drawn from observation. He labels the parts of buildings with letters and identifies them. R was found outside Rome on a bridge over the Tiber. V is above a triumphal arch in Verona. T is in Rome on a Doric temple in the Tullian Prison. "The capital P was found in Pesaro with many other praiseworthy things: its projection, although it may be large, is nevertheless very pleasing to viewers." The podium, base, and capital A are in the Forum Borarium in Rome.[48] This is a collection of architectural fragments, carefully drawn from actual examples, for the use both of practicing architects and students of Vitruvius and of ancient architecture.

It is probable that Vesalius saw Serlio's book while he was working on the *Fabrica* and took some ideas from it. Serlio's architectural representations with their softly shaded renderings resemble stylistically some of Vesalius's carefully depicted body parts. Vesalius himself strove to integrate the artisanal practices of the surgeon and the apothecary into the practice of medicine as a whole. He believed that he was restoring anatomy to its ancient splendor. In his view, medicine was destroyed when its various components such as surgery were broken off from it and relegated "to laymen and people with no knowledge of the disciplines that go to serve the healing art." Similarly, the art of drugs and medicines was handed over to apothecaries. Trained physicians only prescribed medicines and regimes for hidden or internal ailments. As a result, "they shamefully cast aside the foremost and most ancient limb of medicine, the limb that above all is founded . . . on the study of nature."[49] Vesalius beautifully illustrates his belief in the integration of learning and hands-on practice and observation in anatomy in his portrait of himself, his own hands holding the partially dissected arm of a male figure, pen, ink, and paper on the table, the latter containing a sentence that replicates the opening words of his chapter on the muscles of the hand. The cadaver's resemblance to the Christ figure and the classical Ionic column in the background suggest that Christian and classical ideals informed Vesalius's interest in hands-on dissection and observation.[50]

2.13. Andreas Vesalius, *De humani corporis fabrica libri septem* (1543), frontispiece. Portrait of author dissecting a hand. Courtesy Rosenwald Collection, Library of Congress, Washington, D.C.

Vesalius emphasizes that the order of books in the *Fabrica* is the same as the order that he has followed during his own dissections in the company of the eminent men of the city. "This means," he notes, "that those who were present at my dissections will have notes of what I demonstrated and will be able with greater ease to demonstrate anatomy to others." Nevertheless, Vesalius believes that his books will be "particularly useful also for those who cannot see the real thing." These individuals will be able, from Vesalius's treatise, to study each part of the body, "its position, shape, size, substance, connection with other parts, use, function and many similar matters." In sum, from his illustrated treatise, they will be able to learn all the things they could study during a dissection. In a remarkable defense of virtual witnessing, Vesalius tells his readers, "Pictures of all the parts are incorporated into the text of the discourse, so as virtually to set a dissected body before the eyes of students of the works of Nature."[51] Just as the architect can study the elements of the building by perusing Serlio's treatise, so the student of anatomy can study human anatomical parts by scrutinizing the illustrations and text of the *Fabrica*.

Vesalius is aware of the opinions of some "who strongly deny that even the most exquisite delineations of plants and of parts of the human body should be set before students of the natural world; they take the view that these things should be learned, not from pictures but from careful dissection and examination of actual objects." He concedes that he would never urge students to use the pictures alone without dissecting cadavers: "Rather, I would, as Galen did, urge students of medicine by every means at my command to undertake dissections with their own hands." Nevertheless, he defends visual representation as a

Ab ossium substantia & constitutione.

qua tantum ex parte, ut femur, in crustentur. quæ item in cartilaginem degeneret: ut nasi ossa, costæ, pectoris os. His igitur differentijs in suum locum asseruatis, eas quæ à substantia & ossium constitutione colliguntur, sermoni hic adijciamus. Sunt enim quædam ossa plane solida, neque quouis pacto effracta, interius cauernulis sinibus'ue obsita uidentur: ut, præter cætera, duo nasi ossa, & quod in oculorum sede minimum, ac superioris maxillæ ossium secundum numerabitur: & ea quæ sesami seminis magnitudini comparantur: & duo ossicula auditus organo propria. quæ omnia nisi temporis successu impensé resiccata, nullam omnino cauitatem intus ostendut. Nonnulla uerò exterius, quasi continua crusta, aut lamina obduceretur, solida apparent, sed effracta. quædam paruis tantu sinibus cauernulisq copactissimæ spongiæ, uel læuissimi pumicis foramina quamproxime referentibus, intus oppleta cernuntur, quandam fungi speciem exprimentes: ut in minorum ossium numero, brachialis ossa, & tarsi ossa: in maiorum autem, sacrum os, uertebrarum corpora, pectoris os, calcis os, talus, deinde & uerticis ossa, quædam uerò preter cauernulas, nulla serie aut numero positas, ampliu aliquem & insigniter cauum exigunt sinum, qui solidissima ualidissimaq ossis substantia circundatus in extremis angulis, osseis ueluti lineis interstinguitur. Hunc sinum singula ossa, quibus obtigit, unicum ferè adipiscuntur. Sunt autem huius classis inter minora ossa, pedij, postbrachialis, & digitorum ossa: atque ex his potissimu primæ & secundæ digitorum acies, quantumuis se-

In 1. de Vsu partiu, et lib. Comment. in Hipp. lib. de Fracturis.

cus Galeno Anatomes professorum præcipuo uisum sit, ossa digitorum solida constituenti. Grandiora uerò ossa eiusmodi sinu donata sunt, femur, tibiæ os, humerus, maxilla inferior, quartum superioris maxillæ os, fron-

Quoniam præsentis differentia in figuris, singulis Capitibus quibus primatim os aliquod describitur, præpositis, ita ut reliqua quæ hoc Capite pertractantur, non est obuia, hic apponimus brachij ossis, seu ut eu Orelso dicamus humeri, secundum longitudinem dissecti al teram partem, quæ ipsius capite quod scapulæ articulatur, cauernulas pumicis modo efformatas, et A insignitas comonstrat: quæad modum et squama illis cauternulis obducta, et B notata. C autem inscribitur portio exterioris superficiei istius ossis, hic costpicua. Insuper D sinum comonstrat ampliter cauum, qui solidissime crassaq ossis parte E ac F indicata circundatus, secus dum humeri longitudinem exporrigitur. Ad sinus uerò summum ubi G reponitur, cum ubi H uisitur, ossee occurrunt lineæ, medullam hoc sinu continentam inexentes. Cæterum sub humero os cymbam reserens, ac undecima figura trigesimi tertij Capitis exprimendu, hic uerò per medium dissectum, et I et K insignitum delineauimus, ut ossis apparret substantia pumicis instar constructa. Ac utrunq quidem L, cauernosam fungosamue hu ius ossis notat substantiam. M uerò, squamam, quæ ossis constituit superficiem.

Et undique illi fungose substantie circumducitur. Præterea huic ossi unum osficulorum subiunximus, quæ primo pollicis pedis interno dio coarctata, in secunda trigesimiertij Cap. figura u et u osten dentur. Atq hoc ideo N notatum integre per medium diuisimus, ut aliquod os impensé solidum, et cauernulis penè destitucum, in con spectum quouis pacto ueniret.

2.14. Andreas Vesalius, *De humani corporis fabrica libri septem* (1543), 2. Humerus bone split lengthwise. Courtesy Rosenwald Collection, Library of Congress, Washington, D.C.

way of learning anatomy. "In fact," he says, "illustrations greatly assist the understanding, for they place more clearly before the eyes what the text, no matter how explicitly, describes." In addition, Vesalius insists that his "pictures of the parts of the body" will give particular pleasure to those who do not have the opportunity to dissect real bodies or are too squeamish to do so.[52]

As he gets into the subject of his first book, namely bones and cartilage, Vesalius compares the function of bones to the function of certain elements in constructed things. Bone, he writes, is "the hardest, the driest, the earthiest, and the coldest" of all the constituents of the human body. "God the great Creator of all things" formed its substance to be this way for good reason, "intending it to be like a foundation to the whole body; for in the fabric of the human body bones perform the same function as do walls and beams in houses, poles in tents, and keels and ribs in boats." Vesalius later describes the cartilages that form the larynx as resembling "the beams which form the framework of a country cottage before the thatch, the facings, and the mud are applied. In fact, when the human bones and cartilages are stripped of their flesh and then assembled together there is no better analogy to describe them than that of the framework of a hut which has been raised but not yet finished off with branches or earth."[53]

Vesalius's many illustrations of particular bones are labeled with both numbers and letters, which tie them tightly to his explanatory text. Within a discussion of the substance of bone, for example, he shows an illustration of a humerus bone split lengthwise. He uses letters and visual indicators to show the nature of the bony substance: the little holes like pumice in the capula are marked A; the scale over these holes is marked B; C shows the outer surface of the bone; D the large hollow space along the length surrounded by the hardest bone E and F. Underneath the illustration of the humerus, Vesalius shows the navicular bone (a small bone of the wrist), which is cut through the middle to show pumice-like bony substance. Finally, at the bottom, N is a tiny bone at the end of the toe, which is cut through the middle to sl. ...has no pumice-like holes at all. This illustration shows Vesalius's extensive cross-referencing as well: he refers the reader to the illustration where the navicular bone can be seen in its entirety.[54]

The careful, beautiful drawings of the *Fabrica* and Vesalius's assiduous cataloging of the bones of each illustration by lists of descriptive identifications marked by letters and numbers make this a treatise in which visual representation and textual description are integrated in a remarkably close fashion. This integration is furthered by the fact that each time a bone or bone part is mentioned in the text, an italic indication is given in the margin showing where the illustration of the bone or part mentioned can be viewed—sometimes chapters away.

Vesalius and Serlio each created new uses for visual representation in their respective disciplines. Both men understood the value of observing the elements of real buildings or real bodies, and both provided detailed illustrations that would allow such observation apart from actual objects by means of detailed drawings. They lived in a world in which artisanal and learned cultures were growing increasingly proximate.

This chapter's focus on art and nature and their relationships with each other underscores the ways in which both entities were cultural concepts that changed from one context to another over the centuries. Ideas concerning art or artisanal skill developed within particular cultural and social contexts in which numerous artisans carried out skilled crafts, from painting and weaving to smelting metals and cultivating crops. Ideas about nature also developed within the context of social and cultural practices, whether the practice of medicine, for example, or that of teaching within the curriculum and scholastic practices of the medieval universities. An important aspect of the changing context was the medieval development of commercial capitalism and the subsequent burgeoning of the production of material objects, including luxury objects and their growing cultural importance. This development entailed expanding urbanism, and the growth of merchant culture. One result was that art and nature came to be thought of as closer together, and even at times interchangeable.

As we have seen, the distinction adopted by many Aristotelian scholars in the medieval period was reinforced by medieval social and occupational practices. The learned Latin study of the natural world (natural philosophy) in the universities was quite separate from artisanal and practical occupations. Artisan/practitioners were trained in apprenticeship arrangements and usually did not know Latin. They produced a vast array of objects, and they carried out practical tasks such as agriculture and navigation. Exceptions existed to this separation, most strikingly within alchemy. To some extent, furthermore, medieval empirical traditions were continued from ancient origins, within such fields as optics and medicine.

On a far broader scale, however, the relationships between nature and the study of nature, on the one hand, and the making of things and productive practices, on the other, significantly changed during the fifteenth and sixteenth centuries. As a result of economic and social developments, art and nature came to be seen as close together. Such proximity increased when craft practitioners and university-educated men—both influenced by humanism—wrote books on the practical and technical arts. One tradition of writing and commentary—the Vitruvian tradition—was particularly influential in bringing together skilled practitioners and university-educated men. It is to this tradition that we now turn.

CHAPTER 3

Artisans, Humanists, and the
De architectura of Vitruvius

Just as the categories of art and nature came at times to be conflated or almost interchangeable from the fifteenth century, the wide divisions between workshop-trained artisans and university-educated scholar/humanists narrowed and in some cases disappeared. Some learned men undertook practices in which they became skilled, and some artisans took up writing, tried to learn Latin, and in one way or another absorbed humanist learning. A few individuals by happenstance were easily able to cross the boundary between manual skill and Latin learning, whereas others struggled to acquire the necessary skills. Substantive communication between the skilled and the learned, both person-to-person and through writing and reading, became increasingly common.

Humanist scholars in the fifteenth and sixteenth centuries pursued an avid interest in the texts and artifacts of antiquity. Workshop-trained artisans joined them in this interest, especially in matters pertaining to ancient buildings and other kinds of objects, such as coins, medals, and sculptures. Skilled artisans studied and measured such artifacts to learn classical modes of expression in order to glean ancient techniques and incorporate them into their own work. Both learned humanists and practitioners eagerly scrutinized ancient texts that shed light on antique artifacts. Undoubtedly the most important of these was *De architectura,* by the Roman architect/engineer Vitruvius. The only fully extant treatise on architecture from the ancient world, it was probably written, or at least completed, in the 20s BCE. In the fifteenth and sixteenth centuries, the *De architectura* was studied intensely, and came to be a text that mediated the worlds of learning and practice.[1]

Vitruvius and the *De architectura*

Vitruvius worked as a military engineer under Julius Caesar in the 40s BCE. He dedicated his *Ten Books on Architecture* to Caesar's adopted son and heir Octavian (the emperor Augustus after 27 BCE), who had given him a pension at the behest of his sister Octavia.[2] The Roman architect treated the forms of temples and other public and private buildings, and the siting and layout of cities, but also topics relevant to construction and to finishing the work, such as building materials, flooring, ceilings, painting, colors, and plaster.[3] Further, he included chapters on matters that are now usually placed under the rubric of engineering—hydrology, water supply (book 8),[4] time-keeping (sundials and water clocks) (book 9),[5] and machines, including cranes and military machines (book 10).[6]

Vitruvius complained about false and unskilled architects who pushed themselves forward to obtain commissions, and he hoped that his own lack of fame and reputation would be remedied by his treatise.[7] In addition to specific details concerning building design, materials, and construction, the *De architectura* treated architecture as a discipline. Architecture, Vitruvius urged, consists of both construction or practice (*fabrica*) and reasoning (*ratiocinatio*). "*Fabrica* is the constant repeated exercise of the hands by which the work is brought to completion in whatever medium is required for the proposed design. *Ratiocinatio*, however, is what can demonstrate and explain to what degree things have been made with skill and reason." Architects who possessed manual skills but no education, he insisted, could not achieve authority commensurate with their labors, while those who put their trust entirely in reasoning and letters follow a shadow rather than reality. Those who have mastered both are fully armed and have arrived more quickly and with authority at their goal.[8]

Elsewhere in his treatise, Vitruvius advises that the architect be literate, know how to draw, know geometry, history, arithmetic, philosophy, and music, be acquainted with medicine, understand the rulings of legal experts, and have a grasp of astronomy.[9] In his emphasis on reasoning as well as skill, Vitruvius undoubtedly was attempting to raise the status of architecture. He reveals that his own education involved both technical and liberal arts. Describing an Athenian law that required parents to educate their children in an art, he thanks his own parents profusely because they had him trained in an art that could not be mastered "without education in letters and comprehensive learning in every field." Crediting the solicitude of his parents as well as his erudite teachers, Vitruvius underscores that he had benefited from both literary (*philologia*) and technical (*philotechnia*) writings.[10]

Vitruvius also viewed architecture as intrinsic to the growth of human civilization itself. Influenced by Lucretius and other accounts, he described the original humans as beasts who lived in caves and forests. The invention of fire prompted these early humans to gather around fires, and this new proximity led them to begin to communicate and acquire language. Then they began to build shelters of various kinds, imitating the nests of swallows with mud and straw, and going on to develop new forms, learning from each other, while also competing to build better houses in a variety of styles.[11]

The *De architectura* transmitted not only many details concerning ancient building design, time-keeping, hydrology, and machines, but also ideals about how the discipline of architecture should be practiced and its central role in advancing human civilization. But in late antiquity, the purview of the discipline seems to have narrowed: Faventinus, a third-century author, wrote a summary of the part of the treatise on private houses and ignored the rest. Yet Vitruvius's entire text seems to have been quite well known in the medieval period. More than eighty medieval copies of the treatise are extant, the earliest a Carolingian exemplar (British Library, MS Harley 2767), written ca. 800. Despite its important medieval presence, the *De architectura* exerted its most profound influence in the fifteenth and sixteenth centuries—not only through its information about the classical canon and ancient building types and forms but also through its ideals concerning the necessity of both reason and fabrication or hands-on practice. Great excitement accompanied Poggio Bracciolini's "rediscovery" of the text in a Swiss Benedictine monastery at St. Gall in 1416.[12]

3.1. Vitruvius, *De architectura libri dece / tr. de latino in vulgare, afficurati, commentate: . . . da Caesare Caesariano* [Como]: G. da Ponte, [1521] fol. 32r. Humans building the first shelters. Courtesy Rosenwald Collection, Library of Congress, Washington, D.C.

In the fifteenth century, interest in the *De architectura* brought together men of learning and workshop-trained practitioners. The difficult Latin text required Latin skills, but its technical detail and obscure references sometimes could be better grasped by someone experienced in analogous practices. The *De architectura* treated ancient building types that were not necessarily familiar to fifteenth-century people. To understand it required the study of ancient buildings, parts of those buildings such as capitals, columns, and architraves, and their measurements. Learned humanists interested in ancient artifacts and practices and builders and engineers who possessed hands-on knowledge of building construction and machines began to communicate and share their respective areas of expertise as a way of understanding both the ancient text and the buildings of antiquity.

Skilled practitioners often struggled to learn Latin, as they themselves wrote treatises on the practices with which they were familiar, often adding classical references and other marks of learning. Men from university backgrounds engaged in reciprocally beneficial substantive conversations with skilled artisans and joined with them to investigate ancient ruins, especially in Rome. The migration of skilled practitioners to learning and writing and of learned and highborn men to an interest in material construction carried out in life the Vitruvian ideal of combining *ratiocinatio* and *fabrica*, although not in a way that Vitruvius could have foreseen. The articulated ideal was repeated again and again in architectural treatises and Vitruvian commentaries throughout the fifteenth and sixteenth centuries.[13]

Brunelleschi, Alberti, and the Rise of the Artisan in Early Fifteenth-Century Florence

Any discussion of architecture and Vitruvianism in the fifteenth century must begin with Filippo Brunelleschi, who first revived classical forms of architecture in Florence. As a young man, Brunelleschi, a goldsmith and architect/engineer, lost the competition for the contract to fabricate relief panels for the doors of the Florentine Baptistery to the goldsmith Lorenzo Ghiberti.[14] Brunelleschi went on to invent artist's perspective and successfully to design and supervise the construction of the immense, double-shelled dome of Santa Maria del Fiore, the Florentine cathedral. Most of what is known about him derives from a biography by Antonio Manetti (1423–1497), a younger contemporary, who was a humanist scholar and the son of a silk merchant. As a young man, Manetti met Brunelleschi near the end of the famed architect's life and wrote his biography in the 1480s, almost forty years after the Florentine master's death. He reports that Brunelleschi's father was a notary and that Brunelleschi himself learned how to read and write at an early age, and that he also knew how to use an abacus and had learned some Latin. Perhaps, Manetti speculates, he was taught Latin because his father thought his son would follow in his footsteps and become a notary. Instead, the son exhibited great interest in drawing and painting, and eventually was apprenticed to a goldsmith.[15] Brunelleschi's dual background may well have allowed him easily to traverse the boundaries between artisanal skill and Latin culture.

Manetti provides a lengthy description of Brunelleschi's trips to Rome, in which he first looked at sculpture, but then began "to give no less time to that order and method which is in the abutments and thrusts of buildings, [their] masses, lines, and *invenzioni* according to and in relation with their function, and to do the same for decorations." He decided, Manetti continues, "to rediscover the fine and highly skilled method of building and the harmonious proportions of the ancients and how they might, without defects, be employed with convenience and economy." Manetti reports that Brunelleschi studied vaults and considered methods of centering them, that he was familiar with contrivances, having made clocks, alarm bells, and devices with springs, and that he thought a great deal about machines for carrying, lifting, and pulling, according to what the exigencies of the situation might be.[16] He, together with the sculptor Donatello (ca. 1386–1466):

> made rough drawings of almost all the buildings in Rome and in many places beyond the walls, with measurements of the widths and heights as far as they were able to ascertain [the latter] by estimation, and also the lengths, etc. In many places they had excavations made in order to see the junctures of the membering of the buildings and their type—whether square, polygonal, completely round, oval, or whatever. When possible they estimated the heights [by measuring] from base to base for the height and similarly [they estimated the heights of] the entablatures and roofs from the foundations. They drew the elevations on strips of parchment graphs with numbers and symbols which Filippo alone understood.[17]

Manetti continues that Brunelleschi spent many years at the work of observing and measuring ancient buildings and that he found many differences in the types of "columns, bases, capitals, architraves, friezes, cornices, and pediments." He also found differences "between the masses of the temples and the diameters of the columns." So, "by means of close observation" he learned to recognize each of the column types—"Ionic,

Doric, Tuscan, Corinthian, and Attic." And, Manetti adds, "He used most of them at the time and place he considered best."[18]

There is some question concerning the reliability of Manetti's account. It is at times highly polemical, especially concerning the baptistery competition and the discussion of Brunelleschi's work on the dome of the cathedral. Manetti reports Brunelleschi's numerous trips to Rome, sometimes taken with Donatello, at times when one or both are documented as having been in Florence. Howard Saalman in his cogent discussion of the reliability of Manetti's account emphasizes that while the Roman trips are completely undocumented, they are not unlikely and that the elderly Brunelleschi may well have discussed them with the young Manetti. In any case, as Saalman notes, "The fact remains that if there had not been any Roman journeys, Manetti would have had to invent them, because they form a significant aspect of the larger scheme

3.2. Florence, Santa Maria del Fiore (the Duomo). A view of the dome. Photo by author.

to which his hero had to conform." That larger scheme was very much influenced by the great Florentine humanist Leon Battista Alberti and by Alberti's treatise on architecture, *De re aedificatoria*. In Saalman's words, by the 1480s, "Any architect worthy of the name simply *had* to have spent some time in Rome, digging and sketching, before starting out on a serious career of building *all'antica*."[19]

Whether or not Brunelleschi actually went to Rome, Manetti's description of his activities there provides a view of approaches used in the investigation of a group of material structures—approaches that were prevalent at least by the 1480s and indeed (as we know from other sources) well before. In Manetti's account, Brunelleschi assiduously examines building structures, noting how the building was constructed and how its forms relate to its functions. He investigates the proportions of structures and carefully measures them. He examines specific parts of buildings such as vaults, and he is reported to be highly interested in devices and machines, including those for lifting and pulling, as well as springs and clocks. Brunelleschi's interest in machines and devices is well known from other sources. He invented unique lifting machinery for the work of constructing the Florentine dome.[20] Manetti's biography suggests, probably accurately, the empirical approaches and values used by Brunelleschi in his investigation of ancient artifacts.

Brunelleschi's younger contemporary Leon Battista Alberti was one of the most prolific and influential humanists of the fifteenth century. Alberti wrote fluently in Latin and Italian not only on literary and ethical topics but also on practical mathematics, painting, sculpture, and architecture. His mathematical works include *Ludi matematici*, a tract on measuring weight, distances, and dimensions; *Elementa picturae,* which describes geometric figures and transposes them onto foreshortened planes; *Descriptio urbis Romae*, which describes a surveying disk that he used to make a measured survey of the city of Rome; *De statua*, which treats human proportions and a device for replicating those proportions in sculptures of human figures; and *Componendis cifris*, which explains his invention of a code wheel. Alberti also designed four churches—Tempio Malatesta in Rimini, San Sebastiano and S. Andrea in Mantua,

3.3. Matteo di Andrea de' Pasti, medal with portrait of Leon Battista Alberti. © Victoria and Albert Museum, London.

and S. Maria Novella in Florence.[21] Further, he designed the palace of the noble Rucellai family, in Florence. He was undoubtedly involved (but in unknown ways) in Pope Nicholas V's urban planning of Rome around 1450[22] and in Pope Pius II's redesign of the Tuscan town of Pienza.[23]

Alberti also painted pictures, although no paintings that are attributed to him with certainty are extant. His immensely influential *Treatise on Painting* explicated the new artist's perspective by a form of projective geometry. Brunelleschi had demonstrated the perspectival projection of the Florentine Baptistery and the Palazzo de' Signori (the present-day Palazzo Vecchio) using a quite different method that involved panels and reflective mirrors. Alberti wrote both an Italian and a Latin version of his treatise on painting, a first draft completed in 1435. Although the Italian version has long been considered a translation of the earlier Latin version, Rocco Sinisgalli has cogently argued that what happened was the reverse, that the Latin was a translation from the Italian. Alberti dedicated the Italian version to Brunelleschi. Martin Kemp rightly notes that Alberti's object in making an Italian version was not merely to provide the text to non-Latin-reading artisans—such as some of those mentioned in the dedication to Brunelleschi: Donatello, Lorenzo Ghiberti, Luca della Robbia (ca. 1399/1400–1482), the sculptor famous for his terracotta glazed roundels, and the painter Masaccio (1401–?1428); but also Alberti here and elsewhere aimed to make Italian as legitimate a learned language as Latin. His dedication to Brunelleschi and his mention of other sculptors and painters signal his great admiration for the visual revolution—grounded in artist's perspective and the adoption of classical forms—that had greeted him when he entered Florence for the first time, perhaps in 1434.[24]

It would be inaccurate to say that Alberti traversed the boundary between Latin learning and artisan skill. Rather, throughout his varied and prolific writings and in his constructive and practical activities, he obliterated that boundary. This is evident particularly in his architectural treatise *De re aedificatoria*, completed circa 1452 and first printed in 1486, fourteen years after his death. The renowned Florentine humanist Angelo Poliziano (1454–1494), in addressing the dedication letter of the printed edition to Lorenzo de Medici (1449–1492), ruler of Florence, praised Alberti's varied and outstanding literary and practical achievements. He wrote: "So thorough had been his examination of the remains of antiquity that he was able to grasp every principle of ancient architecture and renew it by example; his invention was not limited to machinery, lifts, and automata, but also included the wonderful forms of buildings. He had moreover the highest reputation as both painter and sculptor." Poliziano emphasized Alberti's abilities in both literary and practical realms.[25]

In the *De re aedificatoria* Alberti carefully distinguishes the architect from the carpenter: "The carpenter is but an instrument in the hands of the architect." But this does not mean that he separated architectural activity from actual building construction. The architect, he says, "by sure and wonderful reason and method, knows how to devise through his own mind and energy, and to realize by construction, whatever can be most beautifully fitted out for the noble needs of men."[26] This preliminary statement signals four central concerns evident throughout the *De re aedificatoria*. First is the rational side of architecture, the use of "wonderful reason and method" through the architect's "own mind and energy." Second is the realization of these ideas in "actual construction."

Reasoning and construction are consonant with and indeed reflect the *ratiocinatio* and *fabrica* of Vitruvius's treatise. Third, Alberti's interest in aesthetics is evident—"what can be most beautifully fitted out." Also evident, finally, is his concern for the moral or ethical aspect of architecture, its intrinsic relationship to the life of humans, its use in the service of "the most noble needs of men."

The modern discipline of architecture often looks on Alberti's treatise as an important originary text. Modern architects tend to be far more interested in design than in actual construction; construction now usually falls under the purview of building contractors, civil engineers, and the skilled building trades. One result is that the *De re aedificatoria* has often been misconstrued as conceiving architecture as an art mainly concerned with design, with Alberti's distinction between the architect and the carpenter being taken to support this view. But Alberti in no way disregards the material and constructive aspects of architecture. He continues the discussion summarized above by itemizing the material goods that architects contribute to human life: shelters of various kinds, walks, pools, baths, vehicles, timepieces, shrines, temples, tunnels, hydraulic works, drains, river dredging, the construction of harbors and bridges, the building of ships, and the construction of ballistics and the machines of war.[27] He retains the full range of activities set out by Vitruvius, that is, those concerned with what is now considered engineering as well as the design and construction of buildings.

His treatise, divided into ten books, is very much concerned with the material and constructive aspects of architecture. Book 2 treats materials including timber, stone, bricks, lime, and sand. Book 3, which concerns construction, treats foundations and walls, pavements and roofing, arches and vaulting, including details about the preparation of materials. His chapters on particular kinds of building construction such as public works and private houses include much detail on how to prepare materials and how to proceed with construction. In a chapter on ornamentation, he discusses tools and machines such as cranes and lifting machines in detail. Elsewhere he treats roads, highways, and canals.[28] This detail on materials and construction is interspersed with erudite discussions of classical texts, of the role of architecture in civic life, and of aesthetic issues. The fact is that Alberti was neither an architect nor an engineer in the modern senses of those words. He was a learned humanist who had acquired a deep knowledge and interest in the practical and technical arts, who had studied Vitruvius and other ancient writers intensely and had used and transformed the ideas that he found there for his own original synthesis, and for use in his own technical practice.

Lorenzo Ghiberti and Antonio Averlino, called Filarete: From Artisan to Author

The study of Vitruvius and the production of writings influenced by such study were by no means limited to learned humanists such as Alberti. Workshop-trained men soon took up their pens and produced writings influenced by the ancient architect and by Alberti. One of these was the goldsmith Lorenzo Ghiberti. Ghiberti was trained in his stepfather's goldsmith shop, matriculating in the goldsmith's guild, the Arte della Seta, as a goldsmith and metal sculptor in 1409. His success in the 1401 competition organized by the Arte di Calimala (the guild of cloth finishers and merchants in foreign cloth) gave him the contract to make the bronze relief panels for the north doors of the Florentine Baptistery, beating out Brunelleschi and other competitors.[29] This award laid the foundations for Ghiberti's life's work and his growing reputation. He established a large bronze-casting workshop in order to create and cast the low-relief panels for the doors of the baptistery. He completed the first set of doors in 1424, at which time they were set up in the main portal of the baptistery, facing the cathedral façade, while the doors that had been completed by Andrea Pisano (ca. 1290–1348/9) in 1336 were moved to the west portal. The Arte di Calimala then commissioned the Ghiberti workshop to make a third set of doors. They were completed by August 1448, while the framing and gilding was completed by 1452. The new doors were placed in the main portal while Ghiberti's earlier doors were reinstalled on the north side. The new doors contained large almost square panels with scenes from the Old Testament set in strikingly illusionistic perspectival settings—the first visible affirmation of Albertian perspective. Remarkably beautiful and displaying virtuoso skill, they must have seemed to fully justify Ghiberti's proud self-portrait bust on the same door—just as, across the square, a relief portrait of Brunelleschi adorns a wall above his tomb in the cathedral that displayed his spectacular dome.[30]

Yet Ghiberti augmented his virtuoso accomplishments as a sculptor with a very different activity—the study of ancient texts and the composition of his own treatise. In the late 1440s, he gradually retired from his workshop and seems to have devoted himself increasingly to his bookish studies. He is recorded as having borrowed as early as 1430 an ancient Greek illustrated text on machines by the first-century author Athenaeus Mechanicus. Ghiberti's treatise in three books—I commentarii—was well under way by 1447 but remained unfinished at the time of his death in 1455. As Manfred Wundram put it, Ghiberti's Commentarii "are outstanding among 15th-century writings on art, both in their ambition to discuss the development of art from antiquity to modern times and in their humanist belief in the supremacy of ancient art." Book 1 of the Commentaries focuses on ancient art, with many paraphrases of the texts of Pliny and Vitruvius; book 2 provides a history of art up through a detailed itemization of Ghiberti's own works; and book 3, clearly unfinished, consists of a group of notes on disciplines necessary to the sculptor—optics, anatomy, and human proportions, the latter based on Vitruvius. The optical section, which combines optics and artist's perspective, consists of a series of notes from such medieval authors as Alhacen, Avicenna, Witelo, John Peckham, and Roger Bacon.[31]

The Commentarii focuses on Ghiberti's own area of expertise—sculpture—as well as painting. It is based on his study of ancient texts, primarily Pliny's De natura rerum (On the Nature of Things) and Vitruvius's De architectura, and it exists in only one, very imperfect manuscript copy in the Biblioteca Nazionale in Florence. Ghiberti manifestly struggled, sometimes unsuccessfully, to understand the Latin texts of Pliny and Vitruvius. Despite misunderstanding numerous details, he grasped much of the substance of these texts. His method

3.4. Lorenzo Ghiberti. Florence, Italy. Panel from baptistery door, "Gates of Paradise." Story of Jacob and Esau. Museo dell'Opera del Duomo. Photo credit: Scala/Art Resource, N.Y.

3.5. Lorenzo Ghiberti, self-portrait. Florence, Italy. From baptistery door, "Gates of Paradise," Photo credit: Timothy McCarthy, Archive/Art Resource, N.Y.

was to extract sentences and phrases from the ancient texts and then mold them to his own purposes. For example, following Vitruvius (*De arch.* I.I.I–IO), he writes that "it is suitable that the sculptor, also the painter, be instructed in all these liberal arts"—namely, grammar (i.e., skill in reading and writing), geometry, philosophy, medicine, astrology, *prospectiva*, history, *notomia* (i.e., anatomy), *teorica disegno*, and arithmetic.[32] Thus does Ghiberti follow Vitruvius's list—with important changes, adding *prospectiva*, anatomy, and *teorica disegno*, and disregarding music and law.

He paraphrases the Vitruvian dictum concerning the necessity of both *fabrica* and *ratiocinatio*, applying it to painting and sculpture rather than to architecture. "[The art of] sculpture and painting," he says, "consists of knowledge of many disciplines and is ornamented with various teachings, which is the high invention of all the other arts; it is fabrication with certain thoughtfulness which is accomplished through material and reasoning." Ghiberti insists that "sculptors and painters who have contended without letters, although they have trained with the hands, will not be able to complete their work with authority commensurate with their labors, while those who proceed with reason and letters only, have followed the shadow, but not the real thing." Following his restatement of these Vitruvian dictums, Ghiberti discusses ancient bronze statuary and clay modeling at length. He expresses great admiration for ancient naturalism, just as Pliny had.[33]

Ghiberti's large workshop trained quite a number of skilled artisans in bronze casting, probably including the sculptor and architect Antonio Averlino (ca. 1400–ca. 1469), known as Filarete (lover of virtue), although documentation is lacking. Averlino's first known activity is his work in Rome on the relief panels for the bronze doors of the basilica of St. Peter's, a work in progress between 1433 and 1442. After being accused of attempting to steal the head of St. John the Baptist from the monastery of San Silvester in 1447, he was imprisoned, tortured, and after his release though the intervention of the pope, banned from Rome.[34] In 1451 he moved to the court of Francesco Sforza, duke of Milan (1401–1466), where he began a new career as an architect. There he supervised the construction of buildings (often over the protests of local builders), and he composed his illustrated *Trattato di Architettura* (1461–1464). He dedicated his original treatise to Francesco Sforza and another copy (with additions praising the new dedicatee) to the ruler of Florence, Piero de Medici (1472–1503). The first vernacular architectural treatise of the fifteenth century, it centers on the planning and building of an ideal city, Sforzinda. The narrative takes the form of a dialogue in which the architect (Filarete, lover of virtue) acts as a near-equal to the duke, a generous and approving patron. Filarete often discusses the plans and progress of the new city with the duke as he also teaches architecture to the duke's son, Maria Galeazzo Sforza, an avid student who eagerly absorbs his instruction.[35]

It has been shown that Filarete was influenced in the creation of his ideal city by the dialogues of Plato—even though few of them had been translated from the Greek when he wrote his treatise. Specifically, Filarete was influenced by the *Timaeus*, which shows how the demiurge (an artisan-like figure) designed the universe and made humans in the image of that universe; the *Critias*, which discusses two cities, including their planning and architecture; and the *Laws*, which describes the social, political, and legal conditions of an ideal city. Filarete would have been able to learn the relevant details of these dialogues through Francesco da Tolentino, known as Filelfo (1398–1481), a famous humanist and philologist, fluent in both Greek and Latin, who worked in the Sforza court and was his good friend.[36]

Filarete's friendship with a learned humanist shaped his vision of Sforzinda, but he did not abandon the Vitruvian idea that both skill and reason were necessary. He condemned those who possessed skill alone. Those who, he said, "know how to put a stone in lime, daub it with mortar, and think they are excellent masters of architecture" commit errors, because they understand "neither the measures nor the proportions" of things pertaining to architecture. Such men who have blind faith in themselves and believe nothing can be done better are like the blind leading the blind—all end in a ditch because of poor guidance.[37] The architect must thus understand measures and proportions—how to design buildings and correctly lay out a site—but he must also, Filarete makes clear, be involved with all aspects of the construction. He should personally choose the master masons and supervise them while they work. He must be aware of every detail, cognizant of the finances of the building, should pay for supplies and pay the workers judiciously without overpaying. He must have a specially close relationship with an overseer, who must be thoroughly informed of the orders and must daily be given the architect's "desires and measures" so that he can communicate them to the others.[38]

Filarete emphasizes that to carry out these designing, measuring, and supervising functions well, the architect should himself be a skilled practitioner. He should know how to build and decorate various things, that is, he should "understand many skills and be able to demonstrate them with the work of his hand, with rules of measure, proportion, quality, and suitability." Without the skill of making things "with his own hand, he will never know how to show them or to explain them so they will turn out well."[39]

Within Sforzinda, the crafts have a special place. The city includes a school, primarily for the benefit of those from impoverished families, but also for the well-to-do who wish to attend. Masters there would teach letters, law, canon law, rhetoric, and poetry. Filarete also intended (although it was "not so dignified") that "some manual arts should be taught here by their practitioners." The faculty would include a master of painting, a silversmith, masters of carving in both marble and wood, a turner, a master of embroidery, a tailor, a pharmacist, a glassmaker, and a master of clay, "that is of beautiful vases."[40] Filarete concludes that "this will be a thing that will last for eternity and, moreover, a thing that has never been done before." Universities exist, he recognizes, where boarding students pay a certain amount (and receive instruction), but "this only applies to students of letters." Yet, he concludes, "The other crafts are also necessary and noble, for there are good masters in them. Moreover, all intellects are not equal. Thus it will be possible for every mind to be trained."[41]

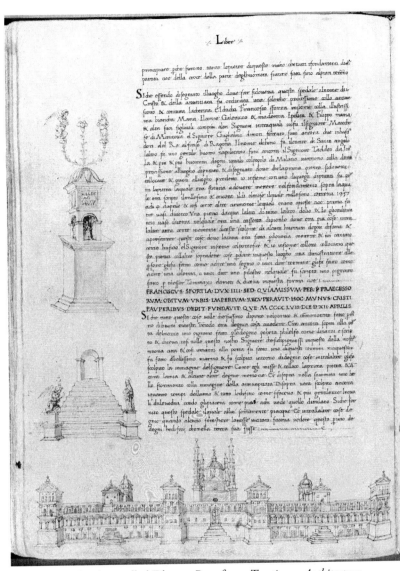

3.6. Antonio Averlino called Filarete. Page from *Treatise on Architecture.* Biblioteca Nazionale Centrale, Florence. Codex Magliabechiano. II I 140, fol. 83v. Photo credit: Scala/Art Resource, N.Y.

In addition to the school, Sforzinda contained what Filarete calls the Houses of Vice and Virtue. The House of Vice accommodates those who enjoy drinking, gambling, and the pleasures of Venus, while the House of Virtue includes rooms for teaching and practicing all the liberal arts, a practice area for military arts, and a theater. Further, "all the crafts or trades that exist are in this place." The House of Virtue includes a temple, open to all, that serves artisans and those who lecture in the arts. Anyone could receive training either in letters or in the arts. Those who were found to be skilled and learned in all the disciplines "were honored for the virtue that they acquired" in their studies. Filarete stresses that all trades are practiced here and that those who are "judged good masters, and if they are young and have been educated in this place, are given the degree like doctors." The House of Virtue was to be governed by three equals—a man of letters, one skilled in arms, and an artisan.[42]

Editions and Translations of Vitruvius

Filarete's attempt to integrate the liberal arts and artisanal craftwork within his ideal city reflected a growing reality for some artisan-trained authors in the late fifteenth and early sixteenth centuries. Francesco di Giorgio, a younger contemporary of Filarete, was well known for his practical expertise in the design and construction of buildings, including fortifications, and as we have seen, he also wrote treatises. Francesco struggled to learn Latin over the years as he also worked on translating Vitruvius. His translation, which exists in one autograph manuscript in the Biblioteca Nazionale in Florence, is the first known translation of the *De architectura* into a vernacular language. Marco Biffi convincingly argues that Francesco prepared it for his own use. Biffi also notes that Francesco's language is typical of an artisan trained in a workshop, and that his Latin, which improved over the years, is that of an autodidact.[43]

"Architecture," Francesco says, "is only a subtle conjecture, conceived in the mind, which is manifest in the work." But, "reason cannot be assigned to each and every thing, because ingenuity [*ingegno*] consists more in the mind and intellect of the architect than in writing or design, and many things happen in the doing [*in fatto*] which the architect or worker never thought of." It is necessary that the architect be both practiced (*practico*) and knowing (*sciente*), that he have a good memory, that he has read and seen many things, and that he be prepared. Thereby he would not resemble "the arrogant and presumptuous" people who are "instructed in errors" and, by demonstrating false things "through force of language," have "corrupted the world." The architect, in contrast, must possess both fabrica (*frabica* [*sic*]) and ratiocination (*raciocinazio* [*sic*]). Following Vitruvius, Francesco itemizes the disciplines in which the architect should be knowledgeable.[44]

Francesco's growing reputation propelled him into a peripatetic career throughout Italy. In the late 1470s he was working in the court of Federico I Montefeltre in Urbino during the same time that the learned philologist Giovanni Sulpizio (or Sulpicius) da Verola (fl. 1470s and 1480s) also was there. Although I know of no documentation that confirms it, it is hard to imagine that these two men would not have been acquainted, and assuming that they were, they surely would have communicated concerning their shared interest in Vitruvius. Sulpizio would edit the first printed edition of the *De architectura*, a Latin edition published in Rome some time between 1486 and 1492. Francesco's decision to undertake a new version of his own treatise (what would become *Trattato II*) was undoubtedly influenced by this printed Vitruvian edition and also by the publication of the first printed edition of Alberti's *De re aedificatoria* in 1486.[45]

Sulpizio taught grammar at the University of Perugia between 1470 and 1475, spent time in Urbino, as mentioned, and was in Rome by 1480. There, he joined a circle of humanists and artists at the Roman academy situated below the Quirinal Hill near the Trevi Fountain and headed by the humanist antiquarian Pomponio Leto (1428–1498). Sulpizio, as professor of grammar, and Leto, as professor of rhetoric, both taught at the University of Rome (the *studium urbis*). Leto avidly instructed his students in the new humanist fields of archaeology, epigraphy, and numismatics, encouraging them to compare ancient texts with ancient ruins and artifacts.[46]

While in Rome, Sulpizio worked on his Vitruvian edition, dedicating it to Raffaele Riaro (1461–1521), a young and energetic cardinal who participated in the circle of humanists around Pomponio Leto. Riaro was engrossed in the planning stages of building his great palace near the center of Rome, now called the Palazzo della Cancelleria. The young cardinal also interested himself particularly in ancient drama and theater design, including Vitruvius's discussion of theaters. He underwrote the first production of an ancient play in Renaissance Rome, Seneca's tragedy *Hippolytus*, or, as it is now called, *Phaedra*. It was directed by Sulpizio and enacted by the professor's students in a temporary theater that the group had constructed in front of Cardinal Riaro's pre-Cancelleria residence, also near the Campo di Fiore. The show was a resounding success, despite the collapse of the set midway through the production, and its emergency repair by Sulpizio and his students.[47]

Sulpizio was intensely involved in a broadly based, collaborative endeavor to study and rediscover Roman antiquity in all its aspects. As Ingrid Rowland has noted, he thought of his Vitruvian edition as a work in progress that would be added to and made more understandable by others in the future. He also addressed a broad readership well beyond elite cardinals such as Riaro. Although the pages of the new edition were not illustrated, they did contain wide margins, the easier to write notes and make drawings on. Sulpizio emphasizes that he undertook the work of editing the difficult text when he observed that if the *De architectura* were widely distributed, it could be very useful, "not only for the learned, but to the rest of men." Indeed, a printed edition was far less expensive than any manuscript could be, and it gave architect/engineers and other practitioners far greater access to the ancient text.[48]

Sulpizio's wish that his edition be studied and improved was at least partially fulfilled. One piece of evidence is a manuscript in the Biblioteca Ariostea in Ferrara that includes a transcription of the *De architectura*, taken from one or both of the slightly revised editions of Sulpizio's Vitruvius that were published in Florence in 1496 and in Venice in 1497. The Ferrara manuscript contains no attribution, but is thought to have been transcribed (with occasional Latin corrections) by the learned humanist Pellegrino Prisciani (ca. 1435–1518). Prisciani worked for Prince Ercole I Este (1431–1505) and assisted the prince in his architectural projects during a time of expanding urbanism and building construction in Ferrara. The manuscript is richly illustrated with drawings—127 of the 196 pages contain illustrations—that seem to have been executed by different hands. The editor of the facsimile edition, Claudio Sgarbi, cogently argues that the text was made for private study and that it and the illustrations may have been used in both public and private lectures and discussions on Vitruvius and ancient architecture. Although precise information is lacking, the unique manuscript seems to have been a collaborative effort of learned humanists and artisans within a princely court led by a prince who was an enthusiastic builder.[49]

Another piece of evidence for the extended use of Sulpizio's edition is a copy of the printed book in the Corsini Library in Rome. It belonged to a Florentine architect who was a member of a famous family of architects active in Rome in the first half of the sixteenth century—the Sangallos. The owner was Giovanni Battista da Sangallo (1496–1548), "il Gobbo"—the hunchback. Rather than leaving the wide margins blank, he filled them with corrections, annotations, translations into Italian, and more than eighty pages of illustrations.[50]

The first illustrated edition of *De architectura* that was printed was produced by a remarkable architect/engineer and learned humanist, Giovanni Giocondo (1433–1515). Although we know nothing of Giocondo's early life or training, his life's work reveals a man highly accomplished both in practical engineering projects such as hydraulics, and in humanist studies. Giocondo worked in the court of Alfonso, duke of Calabria (later Alfonso II, king of Naples), between 1589 and 1591. He was a diligent student of Vitruvius, and he extensively investigated ancient ruins in Naples and in surrounding towns such as Capua and Pozzuola. After 1590, at the death of the architect Giuliano da Maiano (1432–1490), he was given a hand in supervising the construction of the Poggio Reale, a villa with vast terraced gardens containing fountains and extensive hydraulic works, overlooking the Bay of Naples. (The villa was destroyed in the eighteenth century.) Giocondo met Francesco di

Giorgio in Naples—Francesco is first documented there in 1491, working for Alfonso in charge of building a fortress. (Francesco contributed to the invention of a new kind of bastion fortification that was effective against gunpowder artillery, and during his lifetime he was involved in numerous projects to build, reconstruct, and redesign forts in various locations in Italy.) Contact between Francesco and Giocondo is certain, because on 30 June 1492 Giocondo was paid by the Neapolitan treasury for 126 drawings on paper to illustrate two treatises by Francesco di Giorgio (treatises that are not extant).[51] The two men had much substantive knowledge to share. Giocondo's later work on fortification in the Veneto (after 1506) suggests the possibility that Francesco taught him much about the topic, and Giocondo's humanist instruction may have helped Francesco on the more erudite second version of his treatise, *Trattato II*.

Giocondo was born in Verona but seems to have spent a substantial amount of time in Rome during his youth. Like Sulpizio, he was part of the informal academy led by Pomponio Leto. Between 1478 and 1484 he was intensely engaged in an investigation of ancient Roman inscriptions that he found on ruined buildings, walls, pavements, bridges, and towers, and in public and private collections in Rome. He created three collections of inscriptions, each somewhat different, the first dedicated to Lorenzo de Medici of Florence. Giocondo, as described by his biographer, had "a scientific and precise knowledge of antiquities by means of true and exact campaigns of measurement, with drawings taken from life and measurements quoted." The first version of his epigraph collection, dedicated to Lorenzo, contained only inscriptions that he personally had seen, omitting those that he found reported in the writings of others.[52] His interest in practical mathematics, measurement, and measuring instruments is evident in a significant number of extant notes and drawings of measuring instruments—a dioptra similar to an astrolabe, an astrolabe, useful for measuring distances, and a quadrant.[53]

Giocondo's varied activities and travels involved him in both practical and literary tasks. He was in France by 1495 (where he traveled in the aftermath of the French king Charles VIII's successful military campaign against Naples). He remained in France about ten years, involved in the reconstruction in Paris of the Pont Notre Dame (destroyed by fire in 1499) and in the construction of fountains at the Château of Blois. He gave public lectures on Vitruvius. He moved to Venice and worked there and in the surrounding Veneto from 1506, involved in the construction of bridges and canals, and then he contributed to the construction of fortifications in Padua and Treviso. Pope Leo X invited him to Rome, where in 1514, despite his eighty-one years, he was appointed architect of St. Peter's, where he served along with Raphael and Giuliano da Sangallo.[54]

Giocondo's edition of Vitruvius, published in 1511, was illustrated with 136 of his own woodcuts and dedicated to Pope Julius II. A reprint edition in a smaller format, published in 1513 with the treatise on aqueducts by the Roman author Frontinus, was dedicated to Giuliano de Medici, brother of Pope Leo X. In this dedication, Giocondo assures readers that there would now be "a sound, useful, and also delightful text, which would provide knowledge of ancient building, clocks, and machines." Throughout the text, he had worked to make the edition useful to practicing engineers and architects, and he included a useful glossary of technical terms and a table of mathematical symbols. "May the artisan prosper," he enthused in his "Letter to the Reader," "and may he add as many liberal studies as correspondingly are as lively froth to the substance."[55]

Giovanni Giocondo was an individual who cannot be classified as solely a skilled architect/engineer or solely a scholar. Although his background training is unknown, his lifetime activity makes it clear that he absolutely falls into both camps, making the distinction irrelevant. His illustrated edition of 1511 influenced numerous further editions, translations, commentaries, and freestanding treatises in the sixteenth century, many of them de facto collaborative efforts between artisan/practitioners and learned humanists (again with the qualification that the two categories increasingly did not exclude each other).

Shortly after the 1513 Giocondo edition appeared, the painter Raphael, who lived in an elegant palazzo in Via Julia in Rome, invited the learned philologist Marco Fabio Calvo (ca. 1450–1527) to live in his house and produce an Italian translation of *De architectura*. Calvo accomplished this, using Giocondo's edition, probably between 1512 and 1516. A copy contains extensive notes and drawings by Raphael himself. In addition to his work on Vitruvius, Calvo created an edition of the writings of the ancient medical author Hippocrates, and he translated into Latin the ancient physician Galen's commentary on the *Epidemiorum* of Hippocrates. Further, he and Raphael engaged in a collaborative project of graphically reconstructing ancient Rome. Raphael was in the midst of creating surveys of ancient monuments and topography for the purpose of making maps. His method involved "plotting points on paper from readings obtained from a surveying instrument kept at a constant orientation by a magnetic compass." At the same time, Calvo studied ancient Roman topographical writings. The project was cut off by Raphael's premature death in 1520. Calvo himself perished in the

a i arca i ſtipi-
tibus robu-
ſteis & cate-
nis incluſa.

3.7. Giovanni Giocondo, ed., *M. Vitruvius per Iocundum solito castigatior factus cum figures et tabula ut iam legi et intelligi posit . . .* (Venice: Ioannis de Tridino alias Tacuino, 1511), fol. 55v. An illustration on building harbors and other structures in water, illustrating Vitruvius's *De architectura*, 5.12. Reproduced by permission of The Huntington Library, San Marino, California.

terrible Sack of Rome of 1527. He was thrown into prison and died of hunger because he was unable to pay the extortionate tribute money demanded.[56]

Publication of the first printed translation (into Italian) and commentary of the *De architectura* occurred in 1521, primarily the work of Cesare Cesariano (1475–1543). Cesariano includes much autobiographical material in his illustrated commentary, revealing his troubled life, but also his valuation of both learning and artisanal skill and the ways that they related one to the other. His father, he recalled, had "made him recite from the little grammatical work of Donato" (a Latin grammar) when he was four and a half years old. The father, who was in the service of the duke Galeazo Sforza (1469–1494), died soon after (in 1480). When Cesariano reached the age of fifteen, he was forced to flee from home because of the "innate violence" of the man his "stepmother" (*noverca*—actually his mother, Elisabetta de Grittis) had married. Cesariano was a student of Bramante before the master left for Rome in 1499. He worked in Ferrara at about the time that the Ferrara Vitruvius, discussed above, was created (during the reign of Ercole I Este). There he became an assistant to Antonio Visconti, an avid student of philosophy, mathematics, and cosmography, for whom he traced diagrams and scientific drawings. In 1500 he married in the city of Reggio Emilia, where he is recorded as a citizen in 1503 and where he worked on frescoes in the baptistery of St. John the Evangelist and in the Palazzo del Podestà. After killing a man in 1507, he was forced to flee. He went to Parma, where he painted frescoes in the sacristy of the monastery of St. John the Evangelist. These frescoes, securely attributed to him, reveal "the geometric discipline derived from Bramante." He returned to Milan circa 1513, where he worked on frescoes and on architectural projects, including the defenses of the ducal palace. In 1518 he is documented as being in Asti in the Piedmont region of northern Italy working on hydraulic engineering related to containing the Tanaro River.[57]

During these itinerant years engaged in painting, architecture, and engineering, Cesariano also worked on his Vitruvian translation and commentary. With his daily earnings, he reports, he "conversed and studied much in order to observe and understand diverse talents and practices of men." In view of his own experience, he took exception to Aristotle's view that the needy cannot devote themselves to good and are lacking in zeal, and that "it is impossible that the poorly born can do good." Cesare detailed his own troubles and included an allegorical illustration of his own life, asserting that he himself had been specially created and educated so he could be "the explicator of this divine work" [i.e., the Vitruvian edition] to the "great utility and necessity" of the world.[58] Pointing to Vitruvius's lament that some unskilled artists were rewarded greatly while the more skilled received little remuneration (De architectura 3.pref.1–3), Cesariano added that we should be scouts for the good works of others, "especially [works] of the educated and wise learned" and "of those artisans who have labored a great deal," who should be rewarded "according to their good, useful, and necessary profession." Artisans should not be deprived of support, "although sometimes fortune may have wearied and harassed them," forcing them into poverty.[59]

Commenting on the relevant Vitruvian passage, he reiterated that "not only architecture, but every other art," is made up "of the work or fabrication and the reasoning." Reasoning concerns the "well-calculated and considered" rational aspects of each art, and involves general rules. The work itself constitutes the particular application. It is necessary both "to know how to say and how to do," and the work—the doing— is "almost of greater necessity" than the rational part, the saying. The rational aspect is "the speaking with reason about the handmade thing" and involves the demonstration of the object from section to section. Further, Cesariano explained that the treatment of materials is the "drawing out of the sense of the thing through explanation, as does the skilled teacher of some technical skill, who demonstrates not only with words, but with actions, in order to teach the uneducated workers." The ability to understand an object is associated with the ability to handle it skillfully. Nothing arises "in this life except as a result of handling." Those people "who know how to work through handling things, themselves give shape to elegance . . . in order to be recognized for their knowledge."

3.8. Vitruvius, *De architectura libri dece / tr. de latino in vulgare, afficurati, commentate: . . . da Caesare Caesariano* [Como]: G. da Ponte, [1521], fol. 92r. Cesare Cesariano's allegory of his own life. Cesariano depicts himself with his back to the viewer and his hand extended to the wheel of fortune on the right. The sign on his back can be translated as: "The learned man in the end is cast forth from poverty." Courtesy Rosenwald Collection, Library of Congress, Washington, D.C.

Thus does the architect derive his knowledge "not only somewhat from teachers, but from nature."[60]

For Cesariano, the actual construction of a material object entailed the demonstration of its rational principles. Beginning a project, "it is suitable to make order so that the things pertinent to whatever we intend to make can be demonstrated." The "conjectures of our thinking are made manifest to us through the order of the subsequent instance or effect. This demonstrates the art and preformation of things through reason with experience." Thus did the material construction of an object demonstrate its rational principles.[61]

3.9. Vitruvius, *De architectura libri dece / tr. de latino in vulgare, afficurati, commentate: . . . da Caesare Caesariano* [Como]: G. da Ponte, [1521], fol. 165r. Machines for lifting. Courtesy Rosenwald Collection, Library of Congress, Washington, D.C.

Elsewhere, discussing machines, Cesariano wrote that one of the meanings of the word "machination" is "the contriving, effecting, and inventing of manual operations." Further, he believed that "this machination [is] intellective, since it is the cause of the formation of crafted instruments, or of artists adept at explaining the effect of whatever we want to complete." So, he continued, "this ingenious mechanical knowledge is necessary not only to the military arts, but to all liberal demonstrations and operations, without which, almost no convenient embellishment of the world would be possible for the use of ordinary life, nor clothing and the countless number of other indispensable fabricated things that are necessary for human use."[62] Cesariano praised the "noble philosophers" who invented machines. They were to be admired for the "understanding contemplation" that preceded their "great knowledge," and for "a burning desire to produce in sensible works with their own hands, that which they have reasoned with the mind."[63] Thus did he thoroughly integrate reason and knowledge with manual fabrication and manipulation and the orderly demonstration of constructed things.

The publication of Cesariano's translation and commentary did not go smoothly. Two sponsors, Agostino Gallo and Aloisio Pirova, undertook to print 1,300 copies, but disagreements soon arose between them and Cesariano. The two men prevented Cesariano from completing his work. When he attempted to leave Como (where the publication was to occur) carrying his materials, he was intercepted, and his papers forcibly taken from him. The editors engaged others to complete the commentary for the last two chapters of book 9 and for all of book 10. Meanwhile, Cesariano traveled to Milan to initiate a lawsuit. Despite these events, he continued to work. A now-lost printed copy once in the Biblioteca Melziana in Milan contained Cesariano's own extensive marginal notes and corrections. More surprising, modern scholars have discovered an autograph manuscript in the National Library of Madrid, which contains Cesariano's commentaries to the end of book 9 and all of book 10, along with thirty-six drawings. Cesariano worked in Milan as an engineer and architect for many years after the publication. He won his lawsuit in 1528 and received a third of the value of the printed copies.[64]

From the early fifteenth century, humanist scholars and workshop-trained artisans studied *De architectura*, copied and translated it, produced editions and translations, discussed it in detail, made drawings referring to its content, created independent treatises that derived directly or indirectly from it, and studied it with reference to actual artifacts in the material world. These men—Brunelleschi, Alberti, Ghiberti, Filarete, Francesco di Giorgio, Sulpizio, Giocondo, and Cesariano, among others—came from quite diverse backgrounds and possessed very different skills. Yet workshop-trained, skilled artisans became scholars and writers, and humanist scholars acquired skill. The two groups had much to communicate with each other and drew closer together. In the face of this development, the strict categories "craftsman" and "scholar" become less and less apropos; such a distinction obscures a more complicated reality. It should be noted that the Zilsel thesis, for both its advocates and its detractors, depended on a strict identification of the two separate categories, "craftsman" and "scholar." This chapter has shown that the terms, insofar as they present a strictly identifiable dichotomy between two separate kinds of people, become distorting lenses when viewed from the vantage point of the fifteenth century.

"Architecture" too can be a distorting lens if viewed strictly from a modern point of view. Although the terms "architect" and "engineer" and their cognates in Latin and various vernacular languages were very much in use in the fifteenth and sixteenth century as they are today, the terms in the earlier period refer to a different range of activity. An "architect" today holds a professional degree involving an agreed-upon course of study and practices with a license that guarantees certain types of knowledge. An "engineer" likewise holds a degree and a license that presupposes a different set of skills and knowledge. Today these terms refer to professionals and a type of professionalization that was absent in the fifteenth and sixteenth centuries. Not only were the terms "architect" and "engineer" often interchangeable in these centuries (despite the growing separate identity of military engineering), but architecture encompassed a far broader range of activities than it does today. It possessed deep ties to the investigation of artifacts, what today would be called archaeology.

It entailed investigative measurement and the study, design, use, and invention of instrumentation, machines, and clocks. Further, architecture tended to be anthropomorphic—the proportions of buildings reflected human proportions that represented the macrocosm—the harmonies of the cosmos itself. Thus was architecture firmly tied to the natural world, including the entire cosmos. The Vitruvian man, appearing in numerous treatises and in Vitruvian commentaries beyond Leonardo's famous exemplar, stands for just this connectedness of architecture to the world in its human and cosmic dimensions.

The Vitruvian tradition, I suggest, served as a catalyst for communication and exchange between learning and skill. Especially in the sixteenth century, the Vitruvian tradition itself continued to serve as a locus of communication, study, and writing. In addition, in many other arenas from arsenals to cities to mines and ore processing, substantive communication developed among the skilled and the learned. Such arenas, which I call "trading zones," became extremely widespread and culturally important in the sixteenth century, and it is to them that we now turn.

CHAPTER 4

Trading Zones
Arenas of Production and Exchange

The study of Vitruvius's *De architectura*, the investigation of ancient artifacts, and the creation of new forms in art and architecture, along with the production of writings concerning these matters, created numerous "trading zones." The metaphor of the trading zone refers to arenas— symbolic or actual places—where people from different backgrounds who might hold quite different views and assumptions communicate in substantive ways. Peter Galison, deriving the idea from anthropological studies, developed the concept for the history of science and applied it to studies of twentieth-century particle physics to explain how subspecialist groups of physicists who took very different approaches to their subject could communicate with each other and with engineers about how to develop particle detectors and radar; and how experimenters, instrument makers, and theorists could communicate without changing their diverse theoretical orientations or practices, while maintaining different ideas about what they were doing and what their results meant.[1]

"Trading zones" as I use the phrase here with reference to fifteenth- and sixteenth-century Europe differs from Galison's meaning in that he deals with highly developed professional groups working within ever-more-specialized subdisciplines. In contrast, the earlier period that is my focus precedes the development of professionalization, especially in areas such as architecture and engineering. There was a kind of fluidity and openness to discussion concerning issues of design and construction and problems in engineering in which a variety of people from diverse backgrounds offered opinions, suggested alternatives, conversed with one another, and produced relevant writings and drawings. What passed for "expertise" could vary from one situation to another and was far more diverse than has been the case since the full development of professionalism (and its requisite educational and licensing requirements) in modern times.[2]

To demarcate "trading zones" more precisely, they must be distinguished from patron/client relationships. In the latter, certainly there is communication or exchange between two individuals, one with greater power and resources than the other. The patron gives money, employment, or some other benefit, while the client provides a service or some kind of compensatory gift such as the dedication of a treatise.[3] Yet unlike in a trading zone, this type of communication does not involve the reciprocal exchange of substantive knowledge or expertise. Likewise, a trading zone is not identical to a gift exchange. In such exchanges, the donor bestows a gift accompanied by the expectation (as Marcel Mauss has shown) of something in return.[4] Such gift items could involve a great variety of objects, including books and natural history specimens. But, in a trading zone, what is traded is substantive knowledge or expertise.

Early modern trading zones consisted of arenas in which the learned taught the skilled, and the skilled taught the learned, and in which the knowledge involved in each arena was valued by both kinds of "traders." (However, this statement must be qualified because increasingly the boundaries between the two became less clear as the two types of peoples acquired each other's skills and practices.) This exchange often involved direct one-to-one oral communication, but it could also involve indirect forms of exchange such as writing a book, which is later read, or the discussing, editing, translating, and commentating of the kind that occurred in the Vitruvian tradition. What was required was that learned individuals valued practical and technical knowledge, not only for what it could achieve in the material world (such as palaces or fine jewelry) but also as a form of knowledge. Similarly, artisan/practitioners valued knowledge of classical texts, archaeology, and other kinds of knowledge traditionally belonging to learned humanists who knew Latin and had received a university education.

I suggest that the number and range of trading zones between the learned and the skilled increased dramatically in fifteenth- and especially sixteenth-century Europe. Within such trading zones, the people "trading" tended to become more like one another and to lose the distinguishing characteristics deriving from their particular backgrounds. Many activities and particular places became trading zones during these two centuries. They include princely courts; print shops that saw extensive collaboration among authors, printers, designers, engravers, woodcutters, copy editors, patrons, and proofreaders; instrument makers' shops; and coffee shops.⁵ In such "trading zones," both practitioners and learned humanists moved closer together in terms of their empirical values, their knowledge base, and their habitual practices having to do with reading and writing, and with designing and fabricating or constructing physical things. Trading zones became middle grounds where learned and skilled individuals interacted and exchanged substantive knowledge as they often also engaged in constructive and productive activities, created innovative technologies, and wrote tracts, pamphlets, and books on the topic at hand. In this chapter I focus on three trading zones—arsenals, mines and metal-processing sites throughout Europe, and engineering projects in late sixteenth-century Rome.

Arsenals: Sites of Innovation and Exchange

Arsenals proliferated throughout Europe in the fifteenth century and expanded in the sixteenth. They became sites for carrying out multiple tasks and for experimentation involving the manufacture of both guns and gunpowder. Men at arsenals tested ballistics, trained gunners, and designed and supervised the construction of fortifications. Some arsenals, including the famed Venetian arsenal, functioned in addition as dockyards in which ships were designed, constructed, and outfitted. The varied activities at arsenals were complemented by a great expansion of writings on artillery and ballistics, fortification, and shipbuilding and other maritime activities.⁶

Testing, precision measurement, and experimentation became necessary aspects of the wide-ranging development of artillery. A late fourteenth-century record exists that shows that the city of Nuremberg had test-fired guns for both the quality of the metal and accuracy before delivering them to the duke of Bavaria. Such testing became standard procedure in arsenals throughout the empire from the fifteenth century through the seventeenth. Empirical practices of gun founding entailed a performance evaluation that involved precision measurement with regard to aim as well as a consideration of metallurgical materials and work methods.[7]

The active experimentation that characterized the development of artillery was evident in many parts of Europe. In the gun foundries of Flanders and Brabant, for example, a series of experiments and inventions brought about the improvement of gun carriages. Gun founders developed a variety of devices that stabilized the gun on the carriage, aided in handling the perennial problem of recoil, and facilitated accurate aiming and firing. Ongoing experimentation also involved the production of gunpowder. By the mid-fifteenth century, the process of corning had been invented. Corned or granulated gunpowder replaced powdered gunpowder, thereby reducing the risk of accidental firing. Another innovation from the mid-fifteenth century created longer cannon, which improved the trajectory of the shot. As the use of artillery expanded, so also did the construction of arsenals. Along with these developments, numerous writings on artillery and gunpowder appeared.[8]

One of the most important arsenals of Europe was created by the Emperor Maximilian I (1459–1519) in Innsbruck in the Tyrol. Mines in nearby Schwaz supplied copper to the gun foundries and silver to pay for them. Large foundries in Innsbruck manufactured guns. Other shops forged, rolled, and beat armor, and fabricated pikes and swords for the infantry. Specialists in workshops in the nearby town of Absam manufactured cannonballs. The great development of the Innsbruck arsenal was in large part the work of the master founder Gregor Löffler (ca. 1490–1564), the first master gunner to become an arms manufacturer. Löffler transformed his foundry in Innsbruck from an artisanal craft workshop to a large industrial plant, a change that met the needs of increasing demand from the mid-sixteenth century.[9]

The Innsbruck arsenal actively experimented and pursued innovations and improvements in the development of artillery, which included ongoing efforts to standardize the caliber of guns. The arsenal also

4.1. Innsbruck Zeughaus, sixteenth century. Bildarchiv der Österreichischen Nationalbibliothek, Vienna (Cod 10816, fol. 2v–3r). Courtesy Bildarchiv der Österreichischen Nationalbibliothek.

designed wheeled carriages for guns. Similar developments occurred in the growing number of arsenals in other parts of Europe. The problem of multiple calibers, which led to inefficiency and slowness of fire in battle, was addressed in many arsenals, for example in England by Henry the VIII (1491–1547) and in France. The French also worked successfully to achieve greater gun mobility and, like the arsenal at Innsbruck, developed a system in which different types of guns retained the same length barrel. Spanish arsenals paid particular attention to light field artillery and small arms. The Spaniards also developed their own unique gun barrels that were widest in the center and double tapered toward the breach and toward the muzzle.[10]

Georg Hartmann (1489–1564), a mathematician and instrument maker, had studied theology and mathematics at the University of Cologne. After a sojourn in Italy, he moved to Nuremberg in south Germany and set up an instrument shop where he made globes, astrolabes, sundials, and quadrants. In 1540 he invented a caliber scale, a metal rule that showed the internal diameter of cannon and the corresponding weights of stone, iron, and lead shot. This instrument made it unnecessary to weigh the shot before loading guns, thus simplifying their use in battle. In England in the same decade, the gun founder Ralph Hogge of Buxted (fl. 1540s) succeeded in casting guns in iron. During the sixteenth century, the kings of Spain were offered and reviewed numerous military inventions—from rapid-firing artillery to transportable bridges, to a portable mill for grinding grain in a fortress under siege. Models, demonstrations, and tests of new devices were commonplace. The crown often referred proposals to the Council of War, which included military experts, for further consideration. Monopolies were granted for devices deemed workable and useful, the most important being an improved match for arquebuses (an early muzzle-loaded firearm), and a new technique of careening, that is, the cleaning and repairing of ship bottoms.[11]

Spanish ship construction in the late sixteenth and seventeenth centuries centered on two basic types of ship. The galley, an oared ship also powered by sails, had a shallow draft and was suitable for use in the Mediterranean. The center for galley construction, especially during the rule of Philip II (1527–1598), was the arsenal of Barcelona. Philip II began the reform of Spanish naval power, and by 1574 had built a fleet of 150 galleys. However, as Spain turned toward the Atlantic, the high-velocity tidal currents, gales, and huge waves made the galley unsuitable. A second type of ship, the galleon, came into use. It was a large three-masted sailing ship suitable for Atlantic seafaring. Either invented by the Venetians around 1520 or developed from the Portuguese caravel (its precise origins are unclear), it was adopted by Spain. Spain's north coast became a center for building galleons. Ongoing discussion, debate, and experiment focused on the best way to build a galleon for stability, maneuverability, and ability to carry sufficient cargo and guns.[12]

In England the ordnance office created during the reign of Henry VIII was located in the Tower of London. Officers of the ordnance included individuals responsible for technical matters. The surveyor was a mathematical practitioner skilled in measurement and surveying, who tested the quality and quantity of armaments and other goods

4.2. A Spanish galleon circa 1580, a type of three-masted ship that came into use in the early sixteenth century and was suitable for Atlantic seafaring. From Edward W. Hobbs, *Sailing Ships at a Glance: A Pictorial Record of the Evolution of the Sailing Ship* (London: Architectural Press, 1925), 59.

when received. He also supervised the proof master's testing of goods and ammunition, and he surveyed the land and building work in the construction of forts. The office also employed engineers who designed and built forts, fire masters (in charge of gunpowder and explosives), master gunners, and ordinary gunners. All needed some degree of mathematical training. Some carried out skilled mathematical practices in the ordnance office as they also disseminated their knowledge by writing books on mathematical and mechanical topics. For example, in the early seventeenth century, the surveyor of the ordnance, Jonas More (1617–1679), pursued wide-ranging interests in practical mathematics and wrote books on mathematics and fortification.[13]

The Venetian arsenal, key to the defense of the Venetian state and to Venetian cultural pride, was famous throughout Europe. By the sixteenth century it had become a vast, multifaceted enterprise. Occupying about twenty hectares of land, the arsenal was surrounded by more than four kilometers of walls and moats. It employed hundreds of artisans called *arsenalotti,* skilled workers who received the only guaranteed wage in Venice. The arsenal was organized to include three largely separate spheres of production. The largest section was devoted to building, repairing, and outfitting ships. Another department manufactured ropes and cables, and a third was charged with the manufacture of arms and gunpowder.[14]

From the early fifteenth century shipbuilders in the Venetian arsenal experimented with a variety of ship designs, often in rivalry with one another. Notable is a dynasty of Greek masters starting with Teodoro Baxon or Bassanus (d. ca. 1407), who brought techniques to the arsenal from the island of Rhodes. Baxon created a number of new designs, including a light galley that he made wider and heavier than the traditional vessel without sacrificing speed. The Venetian Senate, which governed Venice and controlled the arsenal, encouraged Baxon as well as native Venetian shipbuilders to produce innovative designs that were seaworthy. After Baxon's death the Venetian Senate attempted to lure his nephew Nicolò Palopano from the island of Rhodes, and finally succeeded in 1424. Palopano and Bernardo di Bernardo, the foreman of the ship carpenters of the arsenal, began a long rivalry encouraged by the Venetian Senate. It continued until Palopano's death in 1437.[15]

It was within the ambience of the Venetian arsenal during the time of Palopano that the earliest extant treatise on shipbuilding was composed. Its author was Michael of Rhodes (d. 1445), a mariner who created and illustrated his book for the most part in the 1430s. Although he did not work directly for the arsenal, Michael wrote his book in its shadow and was probably assisted, at least with information, by someone inside. Presumably from the island of Rhodes, Michael hired on to a Venetian galley in 1401 in the low position of oarsman, when he was about sixteen years old. Thereafter, he worked his way up into various officer positions in over forty voyages, which he carefully recorded in the autobiographical service record that he wrote into his book. He gave his position on board, as well as the names of the captain and noble patrons of most of the ships on which he served.[16] His book contains an abacus or mathematical treatise of more than two hundred pages, revealing that he was a good mathematician;[17] a portolan (navigational directions),[18] a section on the zodiac with charming illustrations of the

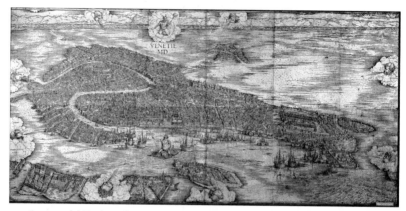

4.3. Jacopo de' Barbari (ca. 1460–ca. 1516), perspective plan of Venice. Museo Correr, Venice, Italy. Photo credit: Scala/Art Resource, N.Y.

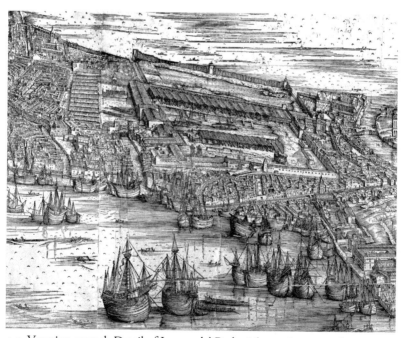

4.4. Venetian arsenal. Detail of Jacopo de' Barbari (ca. 1460–ca. 1516), perspective plan of Venice. Museo Correr, Venice, Italy. Photo credit: Scala/Art Resource, N.Y.

zodiacal signs;[19] and much calendrical material concerning such matters as the date of Easter and the dates of the full moon.[20] He created his own unique coat of arms (arrogating to himself a privilege allowed only to nobles) with a mouse eating a cat perched on top, two turnips on the side, and an M blazed in the middle.[21] The shipbuilding section, which treats the construction of three types of galley and two diverse round ships, contains numerous drawings with measurements, such as those related to the construction of the hull.[22]

Michael probably wrote his book as a way to impress the Venetian nobles who hired officers for their ships for each yearly voyage. Although he was a practitioner—a navigator and mariner—and although his book concerned the practices with which he was involved, Michael's book is not a practical manual; rather it served different cultural uses within wider social spheres within the culture of Venice and the Venetian maritime enterprise. It is a book by a practitioner that, as Piero Falchetta in particular has shown, is a step on the way to luxury navigational books destined for the library shelves of elite merchants and oligarchs. Indeed, Michael's book is evidence of a trading zone. It shows his learning of mathematics, astrology, calendrical matters, and shipbuilding and that this knowledge went beyond the strictly practical aspects of his occupation as a mariner. It seems to have been written with the

4.5. Coat of arms of Michael of Rhodes. Pamela O. Long, David McGee, and Alan M. Stahl, eds., *The Book of Michael of Rhodes: A Fifteenth-Century Mariner's Manuscript*, 3 vols. (Cambridge, Mass.: MIT Press, 2009), 1:329 (fol. 147b). Courtesy MIT Press.

nonpractitioner in mind—elite Venetians who themselves would have been impressed and interested in the practical and technical aspects of shipbuilding and navigation.[23]

Michael in general was an autodidact whose great skill in mathematics suggests that at some point he may have found instruction from one of the many abacus masters who worked in Venice and elsewhere. Probably a native Greek speaker, he wrote (or sometimes copied from other texts) in Venetian. As he worked his way up from the very low position of oarsman, he clearly labored to acquire graphic skill and become knowledgeable in diverse areas. He eventually attained, for some voyages, the highest officer position possible for nonnoble mariners; officers in this position and in some of his other positions were permitted to eat at the captain's table. Whether he actually instructed the young Venetian nobles and other travelers in mathematics, as David McGee has speculated, is unknown.[24]

An intriguing coincidence puts Michael of Rhodes in the same convoy as Nicholas of Cusa (1401-1464), perhaps the greatest philosopher of the fifteenth century. Both men were in the same convoy of four ships sent to fetch the Byzantine emperor John VIII Palaiologos and his party of seven hundred traveling from Constantinople to Venice in 1437. The purpose of the trip was to bring the emperor to the Council of Ferrara-

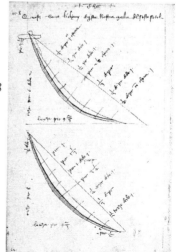

4.6. Diagrams for measuring out the bow and stern of a light galley. Pamela O. Long, David McGee, and Alan M. Stahl, eds., *The Book of Michael of Rhodes: A Fifteenth-Century Mariner's Manuscript,* 3 vols. (Cambridge, Mass.: MIT Press, 2009), 1:348 (fol. 158a). Courtesy MIT Press.

Florence, which many hoped would unite the Catholic and eastern Orthodox churches.[25] Whether the two men were on the same ship and, if so, whether they conversed is unknown. Nevertheless, Cusanus later wrote a treatise, *Idiota: De sapientia, de mente, de staticis experimentis,* in which he advocated the knowledge of the "unlearned" (the *Idiota,* that is, one without knowledge of Latin) and promoted the value of practical mathematics. The first two of the four books take the form of a dialogue between an unlearned man (the *idiota)* and an orator. The *idiota* shows the way to wisdom by rejecting the learning of the orator based on the authority of books. He suggests instead that wisdom can be found in the streets and marketplaces where ordinary weighing and measuring occur.[26] An intriguing, but undocumented, possibility is that the *idiota* in the dialogue could have been modeled on Michael of Rhodes.

Although Michael was an excellent mathematician, he was not a shipbuilder and would not have been able to build a ship. He undoubtedly obtained drawings and other information concerning ship construction from someone, probably one of the Rhodians, working in the arsenal. After the death of the Rhodian master shipbuilder Palopano in 1437, shipbuilders in the Venetian arsenal through the fifteenth and sixteenth centuries continued to produce innovations with, as we have seen, the active involvement of the Venetian Senate. Within this atmosphere of self-conscious rivalry and experimentation, shipwrights created new versions of great merchant galleys, an armed sailing ship called the *barza* (a round ship designed for fighting pirates in the Mediterranean), and light galleys. New ideas often required the presentation of models and arguments in favor of the efficacy of the design over the objections of detractors. This long tradition of naval construction and experimentation at the arsenal provided the ideal setting for Vettor Fausto (after 1480–ca. 1546). Fausto was a humanist who won the position of public lecturer of Greek eloquence in Venice and then embarked on a project to design and then improve on the quinquereme, an oared ship with five rows of oars that had been used by the Greeks in antiquity. Fausto had studied both literature and mathematics and had produced a text and translation of the pseudo-Aristotelian *Mechanics.* The Venetian Senate reviewed his model, and after much debate, provided him with the materials, space, and personnel at the arsenal to build it. Launched in 1529 to great public fanfare, Fausto's quinquereme was taken to be a victory in the revival of Greek science.[27]

Fausto eventually was given a permanent position in the arsenal, and he continued to produce innovations as he directed the construction of ships. Like Alberti, Fra Giocondo, Georg Hartmann, and many others, he is a figure for whom it would not be possible to separate his learning and his technical skill—he seems to have fully possessed both. His influence was still in evidence at the end of the sixteenth century. In 1593, one of his pupils, the shipwright Giovanni di Zaneto, applied some of the principles of the quinquereme to the design of the *galeazza*, a great galley adapted specifically for war. Zaneto's goal was to make this ship as mobile as light galleys. Among the individuals consulted in this matter was the local professor of mathematics, Galileo Galilei (1564–1642). Galileo concerned himself with other military matters as well, such as his invention of a military compass, about which he also wrote a small book. Jürgen Renn and Matteo Valleriani have argued that early in his career Galileo was closely connected to the arsenal and that the development of his thought was strongly influenced by the practical problems, especially concerning the strength of materials, that he confronted there.[28]

Although the book of Michael of Rhodes remained in manuscript form until the twenty-first century, Venice in the sixteenth century functioned as one of the great printing centers of Europe and produced large numbers of practical and technical books, many on topics relevant to the arsenal. They included books on artillery, fortification, and other aspects of the military arts, such as the posthumously published *Pirotechnia* by the metallurgist Vannoccio Biringuccio (1480–ca. 1540), who headed the armory at Rome in his last years. The *Pirotechnia* contained the first detailed description of the casting of bronze cannon and also described boring methods and explained how to produce standard calibers.[29] Niccolò Tartaglia (1500–1557), a mathematics teacher from Brescia who worked in Venice, produced two works on mechanics and mathematics that investigated problems of ballistics: *Nova scientia* (1537) and *Quesiti et inventioni diverse* (1546). Tartaglia uses the dialogue form in his tracts. He depicts himself and others such as gunners discussing a variety of questions with noble princes and dukes. The conversations between gunners and nobles concern topics such as the mathematical trajectories of cannonballs, the best angle for aiming the cannon barrel, and other issues of ballistics. Tartaglia worked on both theoretical and practical mathematical problems. He devised a gunner's quadrant that could help determine the correct position and angle for the efficacious firing of cannon. He also made diagrams of ballistics and analyzed ways of aiming accurately. He produced a table of calibers that mentions twenty-four kinds of guns.[30]

Mines and Ore-Processing Sites

Mining in late medieval Europe was closely associated with arsenals because guns, large and small, were made of metal—either iron or bronze. Eventually cannonballs were made of iron rather than stone. By the mid-fifteenth century the expanded manufacture of gunpowder artillery and the proliferation of princely and noble mints for the production of specie led to a scarcity of metals. Scarcity provided princely and wealthy investors with motivation to take on the cost and technical problems associated with digging deeper mines, and as a result, mining changed radically in the mid-fifteenth century. Medieval mining had usually constituted a local, often family-based, small-scale enterprise, sometimes carried out seasonally as a supplement to agriculture, and it was limited to shallow mines. The new capitalist mining enterprises, in contrast, constituted large-scale operations that employed many workers for wages. These operations profited in part by digging deeper mines and solving the technical problems of ventilation, water removal, and ore removal that accompanied greater depth. To support such endeavors, princes and wealthy entrepreneurs invested money, buying shares in mine operations. For about a hundred years, between 1450 and 1550, they were richly rewarded by a central European mine boom with greatly increased production of silver, copper, iron, tin, and lead.[31]

To solve the problem of removing water and extracting ore from deep mines, miners employed large pumps and other machinery, often powered by waterwheels. Types of water-removing machinery illustrated in Georg Agricola's famous *De re metallica* (1556) include piston pumps made from hollowed-out tree trunks, endless bucket chains, and reversible hydraulic wheels. The most productive mines were in central Europe, in the Erzgebirge mountains, at Schwaz in the Tyrol, and in Hungary. These locations were the sites of large-scale operations such as excavation and processing of silver-bearing copper. Several thousand workers, women as well as men, might work at a single mine; some worked underground; some carried materials; some prepared charcoal; and some were involved in separating, smelting, and refining ores.[32]

The largest ore-processing operations involved the production of silver and copper after the discovery of new methods for processing silver-bearing copper ores. *Saigerhütten*, as they were called in central Germany, were constructed in Saxony, the Tyrol, and elsewhere. These large plants included many hearths, furnaces, bellows, hammers, stamping machinery (most driven by waterwheels), crucibles, and many kinds of tools. Other metal-producing regions included Sweden and Alsace for silver production and Italy for alum, discovered in Civita Vecchia in 1451 and essential in the textile industry for fixing dyes. Iron production expanded rapidly in many areas of Europe in the sixteenth century. Iron processing was transformed by the development of the blast furnace, invented through a process of modifying the traditional bloomery furnace. The bloomery furnace produced a spongy iron called a bloom that was further worked by hammering at a forge. The blast furnace achieved higher heats by means of larger bellows, higher chimneys, and other modifications. Instead of the bloom, it produced molten iron that poured into forms known as pigs. Blast furnaces required greater capital investment and had to be operated continuously for effective production. By the mid-sixteenth century, cast-iron production included products such as cannonballs, pots and pans, and guns. Liège, France, became a

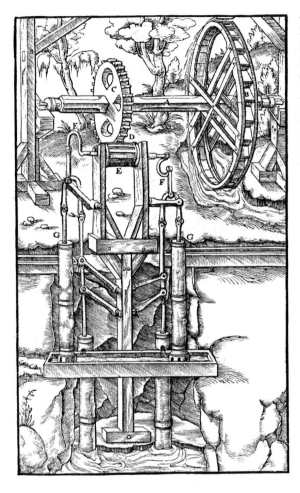

4.7. Piston pumps for removing water from a mine driven by an overshot waterwheel. Georg Agricola, *De re metallica* (Basel: Froben, 1556), 147. Reproduced by permission of The Huntington Library, San Marino, California.

site for a large-scale coal-mining operation. Tin was produced in large quantities in the English counties of Cornwall and Devon.[33]

Technical innovation and mechanization were fundamental hallmarks of early modern mining and ore processing. Mines were dug deeper; galleries and shafts were improved; winches and hoists were installed for ore removal; waterpower increasingly was employed to power pumps and other water-removal equipment, with the more efficient overshot wheel gradually replacing the undershot. Blast furnaces were improved and increasingly used in iron production. Experiments in making alloys and compound metals were ongoing and often entailed the modification of existing techniques. In one innovation, lead was used to make tinplate. Bismuth, discovered in the thirteenth century, was developed commercially for the first time in the sixteenth when it was added to lead and tin to make metal type. Printers' type was later fabricated from harder alloys of lead and antimony, which were also cheaper. Lead was also used in a new technique to make tinplate.[34]

Perhaps the most important innovation came out of a series of experiments that resulted in a technique of processing argentiferous copper ores by alloying copper and lead to produce silver. The three-stage process entailed the addition of lead. First the ore was melted at high temperatures, creating an alloy of lead and argentiferous copper. As the ore cooled, the different temperatures at which copper and lead melt allowed separate crystallizations of copper and silver-bearing lead. This product was then heated in a roasting oven, wherein the lead ore gave off copper crystals. The copper was refined in a drying oven, and the silver was separated from the lead in cupellation ovens. The elements of this complex system developed from alchemical traditions and from the expanding cumulative knowledge acquired from minting coins. It is first documented in Nuremberg in the mid-fifteenth century. It is an invention at the heart of the central European mine boom, producing much higher yields of silver and thereby increasing the profitability of silver and copper mining. It could not be adopted straightforwardly to particular mining operations but often had to be modified to take into account the specific qualities of various local ores. Such modifications required ongoing experimentation. There is evidence that princes were directly involved in initiating experiments by goldsmiths, metallurgists, and others to adapt the process to local conditions in planning mines and ore-processing operations.[35]

Mines constituted important early modern sites for technical experimentation and innovation. Their cultural significance increased with the proliferation of pamphlets and books on mining and metallurgy, especially important from the early sixteenth century. Mine overseers, assayers, and other practitioners, learned humanists, and occasionally nobles wrote books ranging from small pamphlets to detailed and lavishly illustrated treatises on mining, ore processing, assaying, mine organization, and mine law. Of the printed books, an early Italian treatise that became well known in German territories was Biringuccio's *Pirotechnia*, published posthumously in 1540. Far more extensive than previous writings, the *Pirotechnia* treated ores and ore processing, assaying, gold and silver refining, alloys, the art of casting, methods of melting metals, small casting, procedures of working with fires, gunpowder, and fireworks. Biringuccio's extensive textual descriptions of many of these processes were illustrated by woodcuts.[36]

Biringuccio was a practitioner and overseer, but other authors of books on mining and metallurgy were university-trained humanists. Calbus of Freiberg (d. 1523) wrote a small dialogue about mining and ores (*Bergbüchlein*). The well-known humanist and physician Georg Agricola (1494–1555) wrote a dialogue titled *Bermanus* (1530), in which three interlocutors, two physicians, and a mine overseer (Bermanus) discuss regional ores and those mentioned in ancient texts, as they wander in the Erzegebirge. Agricola portrays Bermanus as combining direct observation and experience with knowledge of ancient texts. Agricola's famous treatise on mining and metallurgy, *De re metallica*, was published posthumously in 1556. The humanist Latin treatise contained a defense of mining modeled after the ancient author Columella's defense of agriculture. It also contained rich detail concerning various mine and metallurgical operations and spectacular illustrations of the machines, furnaces, and various processes involved in mining and ore processing. Lazarus Ercker (ca. 1530–1594), an assayer and overseer, wrote a number of books on assaying and ore processing, the last of which was an expansive treatise with illustrations in which the author emphasizes the importance of his own practical experience. Most sixteenth-century mining and metallurgical treatises were printed, but some, such as the *Schwazer Bergbuch*, were written expressly as manuscript books. The copies of the *Schwazer Bergbuch* are adorned with beautifully handpainted miniature illustrations of various mine operations. It was written by Ludwig Lässl (d. 1561), an official in the mine court of Schwaz in the Tyrol.[37]

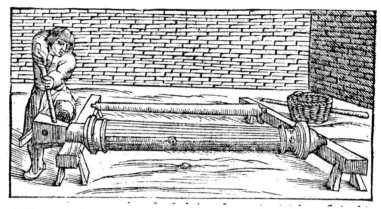

4.8. Vannoccio Biringuccio, *De la pirotechnia* (Venice: Venurino Roffinello. Ad instant a di Curtio Navo and Fratelli, 1540), fol. 83v. Making the barrel and bore of a gun. Reproduced by permission of The Huntington Library, San Marino, California.

Books on mining, ore processing, and metallurgy were written mostly for princes and for a far-flung group of investors in mining, most of whom did not possess mining skills or specialized expertise. The books set out many technical processes in written form, rationalizing the disciplines of mining and metallurgy, using and in part creating precise technical vocabulary. Many of the treatises contained illustrations that provided vivid visual detail of machines and operations. The books often described with great clarity technical operations and equipment. Illustrations were often essential for making complex machinery comprehensible, but they also made the mechanical arts of mining and metallurgy dramatically appealing to the unskilled.

The City—Rome in the Late Sixteenth Century

A very different kind of location that became a trading zone was the city. Cities were expanding in the late sixteenth century, becoming sites of building construction, hydraulic engineering projects, and other projects of urbanization, such as the construction and paving of streets. Here I discuss Rome in the late sixteenth century. As a capital city headed by the pope, Rome cannot be called typical, although it did share some characteristics with other expanding cities of the same period.

Rome in this period became a particularly important center of communication between the skilled and the learned. The urban space was in the process of being transformed by intense activity involving building construction and urban projects—the construction and renovation of churches and palaces; the repair and reconstruction of two great aqueducts, the Acqua Vergine and the newly named Acqua Felice; the creation of new fountains made possible by the greatly augmented water supply; the widening and paving of streets; the redesign of squares; the transport of obelisks from their ancient sites to new locations; and numerous projects aimed at preventing the periodic disastrous flooding of the Tiber River. The goal of such activities was the renewal and transformation of the city.[38]

Especially after the reform of the Catholic Church undertaken after the conclusion of the Council of Trent in the early 1560s, the popes renewed their efforts to create a splendid city consonant with its role as the capital of the Christian world. The intense activity of construction and engineering, combined with the complex patronage situation characteristic of Rome, not only transformed the urban landscape but also brought about the production of numerous writings and images, including cartographical images, related to the practical and engineering concerns of the city. In many urban projects learned individuals interacted with engineers and artisans in significant ways.[39] This extensive interchange between learned and technical cultures could have happened only before the advent of professional engineering and outside of the context of powerful guild control, in a city that was expending construction and engineering efforts in many directions. Here I give two examples of this interchange, in the area of flood control of the Tiber River and in the transport of obelisks as part of the redesign of the streets and plazas.

First, flood control. The Tiber River, which flowed through the center of Rome and provided essential water and other resources to the city, was prone to flooding. The disastrous flood of 1,,, resulted in widespread destruction to the city and to its infrastructures such as sewers and bridges. It was neither the first nor the last flood of the unruly Tiber River, but it was a particularly severe one. Because of its severity and also perhaps because of the particular era in which it occurred, one result was an unusual number of writings by both practitioners and learned men that treated the problem of flooding from a variety of points of view and suggested diverse solutions.[40]

The physician Andrea Bacci spent the years from 1558 to his death in 1600 concerned with the flooding of the Tiber. He wrote his first treatise on the subject in 1558, and then continued to expand and revise it. There is an enlarged revision of 1576; a manuscript tract dedicated to Clement VIII, who became pope in 1592, written in the flyleaves of a copy of the 1576 edition in the Vatican library; and there is another expanded revision of 1599. Bacci had an ongoing interest in natural philosophical issues such as the nature of water and the causes of flooding, which included reflections on Aristotelian texts such as the *Meteorology*. He also provides recommendations for flood prevention in Rome that amount to what he considers a return to ancient Roman river management practices—fortify and augment the banks of the river, lower the riverbed to its ancient level, clean the drains and sewers, and appoint a caretaker of the rivers.[41]

The military engineer Antonio Trevisi arrived in Rome in 1559, and he soon brought out two publications that pertained to flooding. The first, in 1560, treated the flooding of the Tiber River. Trevisi dedicated it to Federico Borromeo (1535–1562), the nephew of Pope Pius IV (r. 1559–1565). He begins his treatise not with the issue at hand, but with a description of the Aristotelian cosmos, the kinds of waters on earth, and the nature of water itself. In these discussions he follows Aristotle and is probably relying on Bacci's treatise published two years before. Trevisi does offer his plan to prevent flooding in one chapter of the treatise but ends with a chapter that consists of a dialogue between a master and an apprentice. Their topic of conversation is not flood control, but how to raise ships from the bottom of a lake. From the mid-fifteenth century the subject had been the focus of intense interest on the part of humanists and engineers, especially because two ancient ships had been discovered at the bottom of Lake Nemi to the south of Rome.[42] Thus did Trevisi present his engineering proposal surrounded by natural philosophical learning, on the one hand, and a dialogue on a topic of humanist and engineering interest, on the other.

Trevisi's second publication was the republication of a map of Rome originally created by a military engineer, Leonardo Bufalini (d. 1552), and first published in 1551. Trevisi's version, published in 1560, seems to have been the same as Bufalini's except that he added letters to the bottom addressed to the Conservators of Rome (the three men who led the communal government), to virtuous architects, to Carlo Borromeo (1538–1584), brother of Federico and nephew to the pope, and to readers. In the letters Trevisi urged his plan for flood control, which involved construction of a huge trench starting below Ponte Milvio and running through Prati (which was then relatively uninhabited) to a low-lying area between the Vatican and Trastevere called the Valle Inferno. Some aspects of Trevisi's proposal were carried out by Pius IV in the early 1560s, including the construction of large trenches around Castel San Angelo, which in part remain.[43]

Other writings on Tiber River flooding included a small tract by the magistrate Luca Peto (1512–1581). Peto was a jurist who wrote the revised law code of Rome, published in 1580. He was also a magistrate who aided the completion of the repair of an ancient aqueduct, the Acqua Vergine, and he wrote a treatise on the weights and measures of the

4.9. Detail from Leonardo Bufalini, *Roma* (Rome: Antonio Blado, 1551 [1560]), showing Leonardo Bufalini and (on the left) his surveying instruments. British Library, Maps S.T.R. [1] © The British Library Board.

ancient Greeks and Romans. Concerning Tiber River flooding, he urged that the number of arches on the bridges be reduced and that the mills on the river be removed.[44]

As the examples of Trevisi, Bacci, and Peto show, writers on flooding of the Tiber dealt with the history of flooding, the natural philosophical issues of the nature of water, and the nature of flooding in diverse areas and offered practical engineering solutions. Whether the authors were practitioners like Trevisi or university-educated men like Bacci and Peto, their writings reveal an interest in both technical and practical issues and in history, natural philosophy, and other aspects of learned culture.

Now I turn to obelisks. The ancient Romans had transported numerous Egyptian obelisks to Rome after the Roman conquest of Egypt following the Battle of Actium in the year 31 BCE. In the sixteenth century, these obelisks stood, or more often lay, in various locations around the city. The most famous was the Vatican obelisk, which had once stood in Nero's circus and now stood in an obscure corner at the side of St. Peter's. Talk had gone on for decades about how to move the giant monolith to the front of the basilica. The difficulty was that the obelisk was immensely heavy at 361 tons and was also extremely fragile. Numerous men put forward technical proposals for the project, accompanied by extensive public discussion. These technical discussions were accompanied by learned discussions of the history of obelisks and the ancient Egyptians and the meaning of the hieroglyphs.[45]

One proposal was written by a fencing master from Milan, Camillo Agrippa. Agrippa had worked on several engineering projects in Rome and also wrote a tract on fencing, another on the generation of the winds and other natural phenomena, a tract on new aids to navigation, and yet another on the way to order troops in battle. These writings point to his broad range of interests, both technical and philosophical. Agrippa also wrote a tract suggesting a method of moving the Vatican obelisk. He was granted an interview with Pope Gregory XIII (r. 1572–1585), which took place in 1582 at the Vatican palace in the presence of the pope's physician Michele Mercati. Agrippa failed to convince the pope that his plan would work. He then wrote a booklet to argue that it was a good plan.

His proposal had the obelisk encased in a protective armature made of thick boards. A high tower would be constructed around the monument consisting of forty timbers leaning toward the top and a platform would be built on the base with rollers underneath. Levers made of oak beams would be suspended half way up and the levers would lift the obelisk. It would be moved on the platform upright and could be lowered onto its new site. At the end of his proposal for moving the obelisk, Agrippa placed a dialogue between two interlocutors on weight and the nature of motion.[46] That is, he moves from technical description to issues of natural philosophy. A very different scheme published in a small booklet

4.10. Portrait of Domenico Fontana holding an obelisk and displaying a gold chain indicating that he had been made Knight of the Golden Spur. Dominico Fontana, *Della trasportatione dell'Obelisco Vatican et delle fabriche di nostro signore Papa Sisto V . . . libro primo* (Rome: Domenico Basa, 1590), frontispiece. Reproduced by permission of The Huntington Library, San Marino, California.

4.11. Moving the Vatican obelisk. Dominico Fontana, *Della
trasportatione dell'Obelisco Vatican et delle fabriche di nostro signore Papa
Sisto V . . . libro primo* (Rome: Domenico Basa, 1590), 18r. Reproduced
by permission of The Huntington Library, San Marino, California.

by one Francesco Masini proposed that canals be built from the site of the obelisk to the new site and that the monument be floated from one to the other.[47]

The architect who actually accomplished the move, Domenico Fontana, wrote a book about the relocation of the obelisk (including spectacular illustrations of the move) and his other architectural and engineering projects. Fontana reports that there were five hundred different proposals for the project. Whether or not this figure is accurate, there must have been numerous proposals, which is an indication of the huge public interest. We know the precise nature only of the final seven proposals, including, in great detail, Fontana's successful one. Fontana's method involved the creation of a *"castello,"* which combined the function of scaffolding, protective armature, and crane. The obelisk was encased in reed mats surrounded by protective planking held together by iron chains. The whole was supported by huge timbers and maneuvered by thick ropes arranged on pulley blocks and pulled by forty capstans, each turned by three or four horses assisted by a number of men and two supervisors. Precise coordination of the capstans was crucial. The obelisk was first lowered, then rolled on a platform, and then raised. The *castello* was taken apart after the lowering operation and reassembled for raising it on its new pedestal in front of St. Peter's.[48]

Before Fontana wrote his own account, Filippo Pigafetta wrote a learned treatise on obelisks, which included their history, a discussion about how to calculate their weight, and a plan for moving the Vatican obelisk (which happened to be Fontana's plan).[49] In 1589 the pope's physician and author of a treatise on mineralogy and metals, Michele Mercati, wrote a learned treatise on obelisks, their history, and the nature of the hieroglyphs, which he believed to be a code for knowledge of the entire natural world transmitted to the Egyptians by Moses. Mercati had been in attendance when Agrippa had described his ideas to Pope Gregory XIII. His own learned treatise also contained a detailed treatment of the engineering project of moving the obelisk that had been carried out by Domenico Fontana.[50]

When Fontana's plan was actually carried out, all of Rome reportedly came to watch. It was a great engineering spectacle that took place in a solemn religious atmosphere. This successful engineering project was executed by one engineer, Domenico Fontana, and one pope, Sixtus V, but also by hundreds of workers laboring in a variety of capacities. It was also tied to numerous writings, public discussions, and treatises written by both practitioners and learned men.[51]

Trading Zones, Elite Individuals, and the Power of the State

Trading zones, as can be seen in the examples discussed here, were closely tied to powerful states. Arsenals were essential to the military defense (and often offensives) of virtually all European states; mining and metallurgy fueled the economies of numerous German states and underlay the political power of the German princes, especially, as well as rulers of other states such as England and Sweden; urbanization in the city of Rome specifically served to enhance the power of the popes. Rulers and powerful elites often exerted strong influence over trading zones and sometimes participated in them personally. Three examples of individuals serve to illustrate this connection. The first is that of the architect Palladio and his wealthy, elite patrons, who eventually became friends and collaborators. The second is Alphonso d'Este of Ferrara and his interest in pottery, and the third, Julius, duke of Braunschweig-Wolfenbüttel (1528–1589) and his attention to mining.

Andrea di Pietro della Gondola (1508–1580), the son of a mason who prepared and installed millstones, was apprenticed to a stonemason in his hometown of Padua at the age of thirteen. He had broken his contract by 1524, and traveled to the small, nearby city of Vicenza, where he joined the guild of masons and stone carvers. Initially he was associated with the workshop of Giovanni da Porlezza, an architect/builder, and Girolamo Pittoni, a sculptor, who together owned the workshop that produced most of the decorative sculpture of the city. About 1537 he met

Gian Giorgio Trissino (1478–1550), a learned humanist, poet, and scholar from a patrician Vicenzan family. They may have met on the worksite in Cricoli outside Vicenza, where Trissino was rebuilding his villa. At the reconstructed villa Trissino established a learned academy for the young noblemen of Vicenza, also asking Andrea, whom he renamed Palladio, to join. Palladio and Trissino developed a long and close friendship, and they traveled to Rome and elsewhere together, studying and measuring ancient buildings.[52]

Palladio's friendship with Trissino would become paradigmatic for other friendships that he made among noble patricians as well as others in the Veneto, as he designed and built many villas in the countryside. He met and became friends with the Paduan humanist and patron Alvise Cornaro (1484–1566), who built a classical-style villa with gardens and wrote on hydraulics, agriculture, and healthy living. Palladio may well have been part of the intellectual circle that met at Sebastiano Serlio's house in Venice before 1541 when Serlio departed for Paris, and he certainly studied and used Serlio's writings. Most important was his friendship with the learned humanist from a patrician family, Daniele Barbaro (1514–1570), patriarch-elect of Aquileia, the Venetian ambassador to England (1549–1550), and superintendent of the University of Padua's new botanical gardens. Barbaro's writings include a treatise on perspective for architects and artists. It is notable that he owned a collection of astronomical instruments.[53]

The collaboration between Palladio and Barbaro began in the early 1550s. They went to Rome in 1554, and worked together on Barbaro's famous Vitruvian commentary of 1556 (with a second edition of 1567), which was illustrated by Palladio. Palladio produced a body of written work as well, most famously the *Four Books of Architecture,* published in 1570, but also other writings, including books on the antiquities and churches of Rome, and an edition of the commentaries of Julius Caesar. Palladio was the one of the designers of the Villa Maser, built for Daniele and his younger brother Marc'Antonio Barbaro (1518–1595), a Venetian senator and ambassador to Constantinople. Marc'Antonio also was an amateur sculptor and especially attended to the fountains and other

4.12. Daniele Barbaro, *I dieci libri dell'architettura, tr. et commentate da monsignor Barbaro* (Venice: F. Marcolini, 1556), frontispiece, *The Measurements of Architecture*. Courtesy Rosenwald Collection, Library of Congress, Washington, D.C.

hydraulic features of the garden at the Villa Maser. The other designer of the villa was probably Daniele Barbaro himself.[54]

Palladio's circle of friends and patrons, including Trissino and Barbaro, was interested in fundamental ways in architecture and building construction, as well as other technical practices. They were also fascinated by Roman antiquities and ruins. Palladio gave friends like Trissino and Barbaro skilled practical knowledge of building design and construction. They gave him access to humanist learning and classical studies.

Interest in hands-on practice on the part of nobles and patrician elites can be found in many arenas beyond architecture and building construction. A striking example—the duke of Ferrara's work at throwing pots—is reported by Cipriano Piccolpasso (1523/24–1579) in his illustrated treatise on pottery. Piccolpasso was one of those midlevel individuals who traversed the boundaries of practice and learning. He was from a poor but noble family originally from Bologna. He was born in Castel Durante (a town famous for its majolica ware) and acquired a humanist education. He followed a varied career, most importantly as *proveditore* in charge of supplies, building materials, and military equipment at the fortress in Perugia, and he wrote a tract on the towns and lands of Umbria, that contained accurately drawn maps of numerous towns in the region. His younger brother Fabio worked as a master in a majolica shop. In his treatise, Piccolpasso provides numerous details on the kiln, potter's wheel, clays, glazes, and pigments, and he gives instructions on how to make majolica ware and pots of all kinds. Piccolpasso reports that the duke of Ferrara, Alfonso d'Este, "took it as his relaxation to have a pottery kiln made for himself in a place near his palace; and thus that wise lord set out of his own accord to experiment concerning these matters, through which he discovered the greatest excellence of the potter's art, yet without laying aside his royal thoughts and his care for his people." Piccolpasso assures us that the making of earthen pots "will not diminish the greatness and worth of so excellent a prince, nor will it obscure the brightness of this White [i.e., the white glaze about to be described]."[55]

4.13. Cipriano Piccolpasso, *Li tre libri dell'arte del vasaio . . . del Cipriano Piccolpassi,* fol. 38v. Preparing colors by pounding them in mortars. National Art Library (Great Britain). Manuscript MSL/1861/7446. © Victoria and Albert Museum, London.

Another territorial ruler, Julius, duke of Braunschweig-Wolfenbüttel, aggressively encouraged the exploitation of mining and ore processing in his territories. Most important were the iron mines and accompanying manufacturing industries, especially those devoted to artillery. Julius experimented extensively on issues concerning artillery and contributed his own inventions. He also opened new mines, expanded old ones, and made administrative reforms to prevent corruption. He sponsored and in part personally wrote and illustrated a book on machines and on ships, *Instrumentenbuch,* an illustrated treatise that exists in one manuscript copy. It depicts machines for removing ores from mines and transporting them. A second section treats ships, locks, water-lifting machines, and other nautical apparatus.[56]

The ties of powerful elites to trading zones were extensive because such zones promoted the economic interests of their states and territories and were important for their personal political and social authority and standing. Arsenals manufactured and maintained the equipment needed for military operations; mines supplied metal for the guns made at the arsenals and produced wealth and economic security; great villas and palaces underscored the political and social authority and legitimation of elite rulers and wealthy oligarchs. Alphonso d'Este, the duke of Ferrara, may have loved the craft of pottery, but it is also true that his territories produced a great quantity of the fine-painted majolica. His interest was tied to the economic well-being of his state. The participation of powerful elites in trading zones had the effect of helping legitimate artisanal, hands-on methodologies and empirical knowledge in general.

The proliferation of trading zones between skilled artisans and learned men (mostly learned humanists) in the sixteenth century helped influence an approach to the investigation of the natural world that valued hands-on experience, accurate measurement, and empirical approaches. This chapter has treated several examples of arenas that became important trading zones—arsenals, mines and ore-processing sites, and a capital city. Such trading zones developed because guns and fortification, mines and metals, and magnificent cities and palaces came to be central to the economic power and political authority of princes and oligarchs.

The development of such trading zones was facilitated by the proliferation of books on technological subjects, books written both by those trained in workshops and those trained in universities. It was in the context both of face-to-face conversations and investigations, and of writing and reading such books, that the skilled acquired learning and the learned acquired skill.

Galison has described trading zones as characterized by the development of pidgin languages and, at times, full-scale Creole languages as a means of communication between people from two different cultures.[57] In the sixteenth century, I suggest, issues of language were mediated by print culture. Books facilitated the entrance of technical vocabulary, utilized heretofore in the oral culture of artisans, into a literate culture of books. Further, as scholastic Latin gave way to humanist forms, some artisans struggled to learn Latin, and at the same time vernacular languages came to be accepted as suitable for learned topics. Much translation occurred as well, from Latin to vernacular and vice versa.

In the sixteenth century, the skilled unlearned acquired some learning, and the unskilled learned acquired some skill. The two groups came closer together. It is inaccurate to call individuals such as Michael of Rhodes, Antonio Averlino called Filarete, Francesco di Giorgo, Leonardo da Vinci, and Palladio only trained artisans. They were also writers, readers, students—indeed scholars. Some learned men such as Giovanni Giocondo and Camillo Agrippa were also highly skilled. These men participated in trading zones as the world of learning adopted empirical values and began to apply them to the investigation of nature.

Empirical Values in a Transitional Age

This book has focused on the thesis that artisans influenced the methodology of the new sciences that developed from the mid-sixteenth century. Marxist scholars such as Hessen, Borkenau, Grossmann, and Zilsel, as well as non-Marxists such as Robert Merton, argued that artisans, or modes of production, or machines used by artisans exerted such influence. The opponents of the thesis, some of them influential figures in the early history of science such as Koyré and A. R. Hall, articulated their opposition in terms of their belief that science was a theoretical enterprise that progressed by advances in theory untouched by the surrounding context. Often left unspoken was the anti-Marxism that also influenced their positions.

Both those who argued for artisanal influence on the new sciences and their opponents accepted without question the rigorously separate categories of "craftsman" and "scholar" that divided makers from thinkers, and in this acceptance both sides joined a long tradition. An important source was Aristotle and Aristotelian traditions that distinguished between *episteme* (theoretical knowledge of the unchanging); *praxis* (knowledge of contingent things requiring judgment); and *techne* (making and thinking about making things). From the time of Aristotle, such categories were hierarchically ranked: *episteme* was at the highest level and *techne* at the lowest. In the medieval period, the liberal arts—the *trivium* (rhetoric, grammar, and logic) and the *quadrivium* (arithmetic, geometry, music, and astronomy)—were considered separate from and superior to the lower mechanical arts.[1] This separation was reinforced in the medieval period by the circumstance that by virtue of background, training, social status, languages used, and place of work, artisans and learned people lived and worked in quite separate arenas.

This book has shown that through the fifteenth and sixteenth centuries practitioners acquired humanist learning and university-educated humanists acquired skill. The two categories overlapped, and the distinction became blurred in certain spheres and arenas. This blurring of two separate realms did not occur universally through all ranks of society, nor did it change the hierarchical social and political structure of that society. Shoemakers and university professors still lived and worked worlds apart in the late sixteenth century, as they had in the twelfth.

The spheres of overlap and interchange between the skilled and the learned did not occur everywhere, but the "trading zones" where they did occur were many and often were situated close to elite individuals and the essential interests of powerful rulers and states. This proximity to centers of power meant that the empirical values promoted in these trading zones gained general currency. In addition, in the two centuries that are the focus of this book, these powerful rulers, princes, and oligarchs caused the built landscape to be visibly transformed with magnificent palaces, churches, public buildings, and redesigned cities. Elites were increasingly surrounded by luxurious goods, ornaments including painting and sculpture in the new style, and lavish apparel. Trading zones framed many of the activities that brought about these changes. As a result, the empirical values characteristic of artisanal culture came to be disseminated widely, making them more readily available as methodological resources for the investigation of the natural world.

My conclusions concerning the development of trading zones are consonant with the interpretive framework of a recent collection of studies, *The Mindful Hand*, in which two of the editors, Lissa Roberts and Simon Schaffer, suggest that employing dichotomous categories such as handwork/intellectual work, craftsman/scholar, and theory/practice distorts the complicated mix of "knowledge, know-how, and technique" (xvii) that characterized early modern European investigations of the natural world.[2] Studies in the volume pertaining to the late sixteenth and seventeenth centuries indeed show in rich detail the complex interrelationships of practical, technical, intellectual, and theoretical practices as they pertain to comets and cannonballs, goldsmithing, and seventeenth-century dioptrics.[3] What I argue is that these close, complex interrelationships were characteristic of the historical period of the fifteenth and sixteenth centuries. Far from being ubiquitous through all time (as Roberts and Schaffer seem to suggest), they are the result of the specific historical developments described here.

By the late sixteenth century, a sharp category distinction between the "scholar" and the "craftsman," the separation of "theory" and "practice," and the distinct categories of "art" and "nature" became anachronistic within certain contexts, despite the continued distinction between the Latin universities and the apprenticeship system of craft training. Developments that brought about the interchange and overlapping of the cultures of learning and of practice include humanist studies outside of the universities, the increasing use of vernacular literatures, the interchange of learning and practice that went on in the courts, the emergence of practitioners who wrote treatises explicating their expertise, and of learned men and highborn rulers who acquired skill in practice; the adoption of values of observation, fabrication, measurement, hands-on practice by university-educated men; and the increased importance and use of instruments, including measuring instruments.

These developments, taken together, point to a long-range cultural change in the fifteenth and sixteenth centuries that led to transformative changes in the ways in which people asked questions about and investigated the natural world. It was not that only "superior artisans" or their products, such as machines, or the organization of their labor influenced the methodologies of the new sciences, as Zilsel and the other Marxists would have it. Rather, it was the interaction of artisans and humanists in trading zones bound by common interests and goals (and the blurring of the differences between them) that brought about profound changes in outlook, changes that favored empirical approaches to investigating buildings, other artifacts, and eventually, the natural world.[4]

The new sciences developed by the efforts of individuals such as Copernicus, Kepler, Galileo, and Newton emerged from the changing culture of fifteenth- and sixteenth-century Europe. This evolving culture brought together the world of manufacture and construction and the world of discourse about the cosmos and the natural world. Such cultural change is evident in many different arenas and locales. It slowly worked to break down and alter a discourse based on a discussion of Aristotelian causes and text-based commentary. Questions beginning with "why," directed at answers involving Aristotelian causation, changed to questions directed at investigations of individual events and objects and their quantitative assessment.

The embrace of empirical values in the late sixteenth and seventeenth centuries made particularly germane those medieval traditions that possessed empirical characteristics. One of these was alchemy. Others included practical mathematics, or "mixed mathematics," such as surveying or observational time-keeping. Medieval alchemical and other empirical traditions may have become a focus of intense interest in the seventeenth century (on the part of Robert Boyle, among others) because they promulgated practices consonant with the empirical values that had gained acceptance in the wider society. It is a distortion to set up an "either/or" distinction—either regarding medieval traditions or changes in the late sixteenth century—in discussing the development of the new sciences. The positive reception of medieval empirical traditions in the seventeenth century undoubtedly reflected the ethos of empiricism that had become a widespread cultural characteristic of late sixteenth-century Europe. What was new about the "new sciences" was not the empirical values of precise measurement, careful observation, the use of instruments, the value of personal experience, and experimentation per se. Such values held wide currency in society at large by the late sixteenth century. What was new was their thoroughgoing application to myriad investigations of the natural world.

The cultural acceptance of empirical values was one of the important foundations upon which the new sciences developed. Such changes by no means brought about an egalitarian society in which craft workers and learned people gained parity. In one sense, the adoption by experimental philosophers of empirical values represents an appropriation.[5] Nevertheless, it is also true that the kinds of practitioners who most intensely participated in trading zones—architect/engineers, painters, and sculptors, for example—saw their own practical disciplines develop new institutional forms, such as the art academies in the seventeenth century, and eventually the professionalization of architecture and engineering in the eighteenth century. The transformation of painting and sculpture into "fine arts" (that have nothing to do with the investigation of the natural world) and the professionalization of architecture and engineering and of "science" itself and its various subspecialties—all are developments of the modern age. This is to underscore that the historical period in which trading zones between workshop-trained artisans and learned humanists flourished was a transitory period encompassing the fifteenth and sixteenth centuries. This particular type of trading zone was distinctly a late medieval or "early modern" phenomenon, not a modern one. Although transitory, it was highly significant as a key to the development of new empirical and mathematical methodologies for investigating the natural world.

Notes

Introduction

1 Thus I use the expression "artisan/practitioners" in a general way to include all skilled workers and practitioners who learned through formal or informal apprenticeships and oral instruction. Although this includes a vast array of diverse skills and disciplines, it points perhaps to a common culture that values handwork and hands-on skill and the practices that accompany them. This is not to suggest that all types of artisan/practitioners exerted influence equally. On the level of particular disciplines or crafts, some practitioners, such as architect/engineers or navigators, as a group, were much more influential than, say, shoemakers or bakers. The influence of artisanal culture as a whole and the influence of particular groups are both important. To complicate matters, as I suggest in this book, the distinctions and separations between certain groups of "artisan/practitioners" and learned men lessened considerably or sometimes disappeared during the two centuries treated here.

2 For a comprehensive introduction to numerous facets of this development, see Katharine Park and Lorraine Daston, eds., *The Cambridge History of Science*, vol. 3: *Early Modern Science* (Cambridge: Cambridge University Press, 2006). Shorter synthetic introductions include Peter Dear, *Revolutionizing the Sciences: European Knowledge and Its Ambitions, 1500–1700* (Princeton, N.J.: Princeton University Press, 2001); John Henry, *The Scientific Revolution and the Origins of Modern Science*, 3d ed. (Basingstoke, Eng.: Palgrave Macmillan, 2008); and Steven Shapin, *The Scientific Revolution* (Chicago: University of Chicago Press, 1996). For a comprehensive historiographic treatment, see H. Floris Cohen, *The Scientific Revolution: A Historiographical Inquiry* (Chicago: University of Chicago Press, 1994); and I. Bernard Cohen, *Revolution in Science* (Cambridge, Mass.: Belknap Press of Harvard University Press, 1985).

3 For a succinct, cogent discussion of the medieval universities and the pedagogical techniques employed therein, see Edward Grant, *The Foundations of Modern Science in the Middle Ages: Their Religious, Institutional, and Intellectual Contexts* (Cambridge: Cambridge University Press, 1996), 33–53.

4 A foundational study is Charles B. Schmitt, *Aristotle and the Renaissance* (Cambridge, Mass.: Harvard University Press, 1983). For Aristotelianism in the medieval universities, see Grant, *Foundations of Modern Science*, 33–53, and see Paul F. Grendler, *The Universities of the Italian Renaissance* (Baltimore: Johns Hopkins University Press, 2002). See also Ann Blair, "Natural Philosophy," in *The Cambridge History of Science*, vol. 3: *Early Modern Science,* ed. Park and Daston, 365–406, esp. 372–379, which emphasizes innovations in Aristotelian natural philosophy during the Renaissance.

5 Although focusing on one city, Richard A. Goldthwaite, *The Economy of Renaissance Florence* (Baltimore: Johns Hopkins University Press, 2009), serves as a comprehensive introduction. For the background, see the classic Robert

S. Lopez, *The Commercial Revolution of the Middle Ages, 950–1350* (Cambridge: Cambridge University Press, 1976), and a recent synthesis, Steven A. Epstein, *An Economic and Social History of Later Medieval Europe, 1000–1500* (Cambridge: Cambridge University Press, 2009). See also S. R. Epstein, *Freedom and Growth: The Rise of States and Markets in Europe, 1300–1750* (New York: Routledge, 2000); and another classic, Fernand Braudel, *The Structures of Everyday Life: Civilization and Capitalism, 15th–18th Century*, trans. Siân Reynolds, 3 vols. (New York: Harper and Row, 1981).

6 See esp. Antonio Barrera-Osorio, *Experiencing Nature: The Spanish American Empire and the Early Scientific Revolution* (Austin: University of Texas Press, 2006); Harold J. Cook, *Matters of Exchange: Commerce, Medicine, and Science in the Dutch Golden Age* (New Haven, Conn.: Yale University Press, 2007); and Marìa M. Portuondo, *Secret Science: Spanish Cosmography and the New World* (Chicago: University of Chicago Press, 2009).

7 See the classic Norbert Elias, *The Court Society*, trans. Edmund Jephcott (New York: Pantheon Books, 1983). In the history of science, a groundbreaking work on the European courts was Mario Biagioli, *Galileo, Courtier: The Practice of Science in the Culture of Absolutism* (Chicago: University of Chicago Press, 1993).

8 See esp. Roger Chartier, *The Order of Books: Readers, Authors, and Libraries in Europe between the Fourteenth and Eighteenth Centuries*, trans. Lydia G. Cochrane (Stanford, Calif.: Stanford University Press, 1994); Elizabeth Eisenstein, *The Printing Press as an Agent of Change: Communications and Cultural Transformations in Early Modern Europe*, 2 vols. (Cambridge: Cambridge University Press, 1979); and Adrian Johns, *The Nature of the Book: Print and Knowledge in the Making* (Chicago: University of Chicago Press, 1998).

9 See Patricia Fortini Brown, *Private Lives in Renaissance Venice: Art, Architecture, and the Family* (New Haven, Conn.: Yale University Press, 2004); and Richard A. Goldthwaite, *The Building of Renaissance Florence: An Economic and Social History* (Baltimore: Johns Hopkins University Press, 1980).

10 See esp. Samuel Y. Edgerton, Jr., *The Renaissance Rediscovery of Linear Perspective* (New York: Basic Books, 1975); Martin Kemp, *The Science of Art: Optical Themes in Western Art from Brunelleschi to Seurat* (New Haven, Conn.: Yale University Press, 1990); and Kim H. Veltman and Kenneth D. Keele, *Linear Perspective and the Visual Dimensions of Science and Art* (Munich: Deutscher Kunstverlag, [?1986]). And see also the important study by Stuart Clark, *Vanities of the Eye: Vision in Early Modern European Culture* (Oxford: Oxford University Press, 2007).

11 William Donahue, "Astronomy," in *The Cambridge History of Science,* vol. 3: *Early Modern Science*, ed. Park and Daston, 562–595. For Copernicus, see especially Robert S. Westman, *The Copernican Question: Prognostication, Skepticism, and the Celestial Order* (Berkeley: University of California Press, 2011). For Brahe, see esp. Victor E. Thoren, *The Lord of Uraniborg: A Biography of Tycho Brahe* (Cambridge: Cambridge University Press, 1990).

12 For a study that delineates the attitudes of numerous artisans, see Pamela H. Smith, *The Body of the Artisan: Art and Experience in the Scientific Revolution* (Chicago: University of Chicago Press, 2004). See also James A. Bennett, "The Mechanical Arts," in *The Cambridge History of Science,* vol. 3: *Early Modern Science,* ed. Park and Daston, 673–695.

13 See esp. Steven A. Epstein, *Wage Labor and Guilds in Medieval Europe* (Chapel Hill: University of North Carolina Press, 1991).

14 Grant, *Foundations of Modern Science,* 33-53. See also Grendler, *Universities of the Italian Renaissance,* 267–313, for natural philosophy, and Pearl Kibre and Nancy G. Siraisi, "The Institutional Setting: The Universities," in *Science in the Middle Ages,* ed. David C. Lindberg (Chicago: University of Chicago Press, 1978), 120–144.

15 For in-depth discussions and an introduction to the large literature on humanism and the issues surrounding it, see esp. Christopher S. Celenza, *The Lost Italian Renaissance: Humanists, Historians, and Latin's Legacy* (Baltimore: Johns Hopkins University Press, 2004); Jill Kraye, ed., *The Cambridge Companion to Renaissance Humanism* (Cambridge: Cambridge University Press, 1996); Albert Rabil, Jr., ed., *Renaissance Humanism: Foundations, Forms, and Legacy,* 3 vols. (Philadelphia: University of Pennsylvania Press, 1988); and Ronald G. Witt, *In the Footsteps of the Ancients: The Origins of Humanism from Lovato to Bruni* (Leiden: Brill, 2000). For humanism in the universities, see esp. Grendler, *Universities of the Italian Renaissance,* 199–248.

16 For writings concerning the crafts by both skilled artisans and learned men, see Pamela O. Long, *Openness, Secrecy, Authorship: Technical Arts and the Culture of Knowledge from Antiquity to the Renaissance* (Baltimore: Johns Hopkins University Press, 2001). For Alberti, see Chapter 3 below.

Chapter 1

1 For a useful overview of the historiography of the scientific revolution as a whole, see H. Floris Cohen, *The Scientific Revolution: A Historiographical Inquiry* (Chicago: University of Chicago Press, 1994).

2 Ibid., esp. 200–204 and 322–327.

3 Karl Marx, *Capital: A Critique of Political Economy,* trans. Ben Fowkes (London: Penguin Books, 1976), 1:452–453, 455–467. I am indebted to Gideon Freudenthal, "Introductory Note," *Science in Context* 1 (1987): 105–108, who discusses Marx's notion of manufacture.

4 My account is indebted to Tom Bottomore, "Austro-Marxism," in *A Dictionary of Marxist Thought,* ed. Tom Bottomore (Cambridge, Mass.: Harvard University Press, 1983), 36–39. See also Bottomore and Patrick Goode, eds. and trans., *Austro-Marxism* (Oxford: Clarendon Press, 1978), which includes a selection of texts by Max Adler and others, translated into English, and an extensive introduction by Bottomore. And see Christoph Butterwegge, *Austromarxismus und Staat: Politiktheorie und Praxis der österreichischen Sozialdemokratie zwischen den beiden*

Weltkriegen (Marburg: Verlag Arbeit und Gesellschaft GmbH, 1991); Peter Heintel, *System und Ideologie: Der Austromarxismus im Spiegel der Philosophie Max Adlers* (Vienna: Oldenbourg, 1967); Norbert Leser, *Zwischen Reformismus und Bolschewismus: Der Austromarxismus als Theorie und Praxis* (Vienna: Europa Verlag, 1968), esp. 511–561; Alfred Pfabigan, *Max Adler: Eine politische Biographie* (Frankfurt: Campus, 1982); and Anson Rabinbach, *The Crisis of Austrian Socialism: From Red Vienna to Civil War, 1927–1934* (Chicago: University of Chicago Press, 1983).

5 See Diederick Raven and Wolfgang Krohn, "Edgar Zilsel: His Life and Work (1891–1944)," in Edgar Zilsel, *The Social Origins of Modern Science*, ed. Diederick Raven, Wolfgang Krohn, and Robert S. Cohen (Dordrecht: Kluwer Academic, 2000), xix–lix, esp. xx–xxvi.

6 A concise summary of the Vienna Circle and its importance to the present day is Thomas Uebel, "Vienna Circle," in *Stanford Encyclopedia of Philosophy,* http://plato.stanford.edu/entries/Vienna-circle/ 28 June 2006, rev. 18 September 2006 (accessed 15 April 2010). See also Friedrich Stadler, "What Is the Vienna Circle? Some Methodological and Historiographical Answers," in *The Vienna Circle and Logical Empiricism: Re-evaluation and Future Perspectives,* ed. Friedrich Stadler (Dordrecht: Kluwer Academic, 2003), xi–xxiii; Friedrich Stadler, "Aspects of the Social Background and Position of the Vienna Circle at the University of Vienna," in *Rediscovering the Forgotten Vienna Circle: Austrian Studies on Otto Neurath and the Vienna Circle*, ed. Thomas E. Uebel (Dordrecht: Kluwer Academic, 1991), 51–77. For the ties between the philosophy of science developed in Vienna at this time (logical positivism, or logical empiricism) and Marxist and socialist political thought and activism, see the astute analysis of Don Howard, "Two Left Turns Make a Right: On the Curious Political Career of North American Philosophy of Science at Midcentury," in *Logical Empiricism in North America,* Minnesota Studies in the Philosophy of Science 18, ed. Gary L. Hardcastle and Alan W. Richardson (Minneapolis: University of Minnesota Press, 2003), 25–93, esp. 28–46.

7 For a history of Vienna during these years, see Helmut Gruber, *Red Vienna: Experiment in Working-Class Culture, 1919-1934* (New York: Oxford University Press, 1991). For Zilsel's life and thought, see esp. Johann Dvořák, *Edgar Zilsel und die Einheit der Erkenntnis* (Vienna: Löcker Verlag, 1981), 33–40 (on education and school reform), 51–63 (on Zilsel and the Vienna Circle), and 63–78 (on the unity of knowledge); and see Raven and Krohn, "Edgar Zilsel: His Life and Work," xix–lix.

8 Believing that knowledge should be unified on the basis of scientific principles, Zilsel was allied with Neurath's campaign to create one unified science through encyclopedism, although he was also critical of Neurath's approach. See also Stadler, "Aspects of the Social Background," and Johann Dvořák, "Otto Neurath and Adult Education: Unity of Science, Materialism and Comprehensive Enlightenment," both in *Rediscovering the Forgotten Vienna Circle,* ed. Uebel, 51–77 and 265–274. And see Christian M. Götz and Thomas Pankratz, "Edgar Zilsels Wirken im Rahmen der wiener Volksbildung

und Lehrerfortbildung," in *Wien-Berlin-Prag: Der Aufstieg der wissenschaftlichen Philosophie: Zentenarien: Rudolf Carnap, Hans Reichenbach, Edgar Zilsel*, ed. Rudolf Haller and Friedrich Stadler (Vienna: Verlag Hölder-Pichler-Tempsky, 1993), 467–473. For an introduction to the life and work of Neurath, see Nancy Cartwright, Jordi Cat, Lola Fleck, and Thomas E. Uebel, *Otto Neurath: Philosophy between Science and Politics* (Cambridge: Cambridge University Press, 1996); and Otto Neurath, *Empiricism and Sociology*, ed. Marie Neurath and Robert S. Cohen, trans. Paul Foulkes and Marie Neurath (Dordrecht: D. Reidel, 1973); and Danilo Zolo, *Reflexive Epistemology: The Philosophical Legacy of Otto Neurath*, trans. David McKie (Dordrecht: Kluwer Academic, 1989).

9 See Nicholas Jardine, "Essay Review: Zilsel's Dilemma," *Annals of Science* 60, no. 1 (2003): 85–94, who cogently explicates some of Zilsel's intellectual conflicts with other Viennese thinkers; and Monica Wulz, "Collective Cognitive Processes around 1930: Edgar Zilsel's Epistemology of Mass Phenomena," http://philsci-archive.pitt.edu/archive/00004740/ (accessed 15 July 2010).

10 Edgar Zilsel, *Die Entstehung des Geniebegriffes: Ein Beitrag zur Ideengeschichte der Antike und des Frühkapitalismus*, preface by H. Maus (1926; Hildesheim: Olms Verlag, 1972) . On Zilsel's ideas concerning the "laws of history," see Zilsel, "Physics and the Problem of Historico-sociological Laws," and "Appendix II: Laws of Nature and Historical Laws," both in Zilsel, *Social Origins*, 200–213 and 233–234. See also Wolfgang Krohn and Diederick Raven, "The 'Zilsel Thesis' in the Context of Edgar Zilsel's Research Programme," *Social Studies of Science* 30 (December 2000): 925–933; Diederick Raven, "Edgar Zilsel's Research Programme: Unity of Science as an Empirical Problem," in *Vienna Circle and Logical Empiricism*, ed. Stadler, 225–234; and Wulz, "Collective Cognitive," 6–8.

11 See Raven and Krohn, "Edgar Zilsel: His Life and Work," xix–lix, esp. xx–xxvi; Diederick Raven, "Edgar Zilsel in America," in *Logical Empiricism in North America*, ed. Hardcastle and Richardson, 129–148; and a moving personal memoir by Paul Zilsel, "Portrait of My Father," *Shmate* 1 (April/May 1982): 12–13.

12 See Dvořak, *Edgar Zilsel*; and Zilsel, *Social Origins of Modern Science*, which includes reprints of his papers on the social origins of the scientific revolution as well as a detailed biographical essay.

13 For Kolman's life and thought, see Pavel Kovaly, "Arnošt Kolman: Portrait of a Marxist-Leninist Philosopher," *Studies in Soviet Thought* 12 (December 1972): 337–366; and see Loren R. Graham, "The Socio-political Roots of Boris Hessen: Soviet Marxism and the History of Science," *Social Studies of Science* 15 (November 1985): 705–722. Graham's view of Hessen has been criticized by Gideon Freudenthal and Peter McLaughlin, "Classical Marxist Historiography of Science: The Hessen-Grossmann Thesis," in *The Social and Economic Roots of the Scientific Revolution: Texts by Boris Hessen and Henryk Grossmann*, ed. Gideon Freudenthal and Peter McLaughlin ([Dordrecht]: Springer, 2009), 1–38, esp. 32–33. For biographical information on Hessen, see Freudenthal and McLaughlin, "Boris Hessen: In Lieu of a Biography," in *Social and Economic Roots*, 253–256. For an early translation of Hessen's 1931 paper, and valuable

introductory material, see Boris Hessen, "The Social and Economic Roots of Newton's 'Principia,'" in *Science at the Crossroads*, 2d ed., introduction by P. G. Werskey (London: Frank Cass, 1971), 149–212.

14 For further elaboration of the situation in the Soviet Union and for communication between Russian scientists and the British, see C. A. J. Chilvers, "The Dilemmas of Seditious Men: The Crowther-Hessen Correspondence in the 1930s," *British Journal for the History of Science* 36 (December 2003): 417–435; and for the great influence of the Russian delegation on a group of British scientists, see Mary Jo Nye, "Re-Reading Bernal: History of Science at the Crossroads in 20th-Century Britain," in *Aurora Torealis: Studies in the History of Science and Ideas in Honor of Tore Fränsmyr*, ed. Marco Beretta, Karl Grandin, and Svante Lindquist (Sagamore Beach, Mass.: Science History Publications, 2008), 235–258; and Jonathan Rée, *Proletarian Philosophers: Problems in Socialist Culture in Britain, 1900–1940* (Oxford: Clarendon Press, 1984). See also A. K. Mayer, "Fatal Mutilations: Educationism and the British Background to the 1931 International Congress for the History of Science and Technology," *History of Science* 40 (December 2002): 445–472, which reconstructs the British context of the meeting as influenced by "educationism," referring to the importance of pedagogical concerns and moral education as central to the developing history of science discipline; and see Vidar Enebakk, "Lilley Revisited: Or Science and Society in the Twentieth Century," *British Journal for the History of Science* 42 (December 2009): 563–593.

15 Boris Hessen, "The Social and Economic Roots of Newton's *Principia*," in *Social and Economic Roots*, ed. Freudenthal and McLaughlin, 41–101, esp. 41–61.

16 Ibid., esp. 73–82. See also Freudenthal, "Introductory Note," 106–107.

17 See Franz Borkenau, "Zur Soziologie des mechanistischen Weltbildes," *Zeitschrift für Sozialforschung* 1, no. 3 (1932): 311–355, translated by Richard W. Hadden as "The Sociology of the Mechanistic World-Picture," *Science in Context* 1 (March 1987): 109–127; and Franz Borkenau, *Der Übergang vom feudalen zum bürgerlichen Weltbild: Studien zur Geschichte der Philosophie der Manufakturperiode* (1934; New York: Arno Press, 1975). For Borkenau's life, see esp. Richard Lowenthal, "In Memoriam Franz Borkenau," *Der Monat* 9 (July 1957): 57–60; Valeria E. Russo, "Profilo di Franz Borkenau," *Rivista di Filosofia* 72 (June 1981): 291–316; John E. Tashjean, "Borkenau: The Rediscovery of a Thinker," *Partisan Review* 51, no. 2 (1984): 289–300; and Tashjean, "Franz Borkenau: A Study of His Social and Political Ideas" (Ph.D. diss., Georgetown University, 1962). For Borkenau's position in the Institute for Social Research in Frankfurt, see Martin Jay, *The Dialectical Imagination: A History of the Frankfurt School and the Institute of Social Research, 1923–1950* (London: Heinemann, 1973), 16–18. A study of another aspect of Borkenau's wide-ranging work is William David Jones, "Toward a Theory of Totalitarianism: Franz Borkenau's *Pareto*," *Journal of the History of Ideas* 53 (July–September 1992): 455–466.

18 For the institute in the 1920s, see esp. Jay, *Dialectical Imagination*, esp. 3–40; Paul Kluke, *Die Stiftungsuniversität Frankfurt am Main, 1914–1932* (Frankfurt am Main: Verlag von Waldemar Kramer, 1972), 486–513; Russo, "Profilo di

Franz Borkenau," esp. 294–299; and Rolf Wiggershaus, *The Frankfurt School: Its History, Theories, and Political Significance*, trans. Michael Robertson (Cambridge:, Mass.: MIT Press, 1994), 9–126. For the intellectual changes of the institute as it transferred to New York, see Martin Jay, *Permanent Exiles: Essays on the Intellectual Migration from Germany to America* (New York: Columbia University Press, 1986), esp. 28–61.

19 For Lukács's influence on Borkenau, see especially Tashjean, "Borkenau," 294–295. For Lukács's concept of reification, see Georg Lukács, *History and Class Consciousness: Studies in Marxist Dialectics*, trans. Rodney Livingstone (Cambridge, Mass.: MIT Press, [1971]), 83–211. See also Gajo Petrović, "Reification," in *Dictionary of Marxist Thought*, ed. Bottomore, 411–413.

20 Yet as John Tashjean and others have pointed out, Borkenau did not posit a simple one-to-one relationship between the labor of manufacture and the scientific world view. Rather, using Lukács's notion of reification, Borkenau elaborated how the forms of labor and the view of the world as a machine, or the mechanistic world picture, developed and reciprocally reinforced one another in the seventeenth century, mediated by the rise of the industrial bourgeoisie. See Borkenau, "Sociology of the Mechanistic World-Picture"; Tashjean, "Borkenau," 294–297; and Jay, *Dialectical Imagination*, 16.

21 Note that Grossmann was spelled Grossman in Poland and North America. See Rick Kuhn, *Henryk Grossman and the Recovery of Marxism* (Urbana: University of Illinois Press, 2007), 1–72 (early years) and 164–167 (for his work on a critique of Borkenau instigated by Horkheimer); Valeria E. Russo, "Henryk Grossmann and Franz Borkenau: A Bio-Bibliography," *Science in Context* 1 (March 1987): 181–191; Jay, *Dialectical Imagination*, 16–19; Jay, *Permanent Exiles*, 29–30; and Martin Jay, *Marxism and Totality: The Adventures of a Concept from Lukács to Habermas* (Berkeley: University of California Press, 1984), 196–219, for the disenchantment of the younger scholars led by Horkheimer and the institute's changed direction. For a detailed overview of these early years of the institute, see Wiggershaus, *Frankfurt School*, 9–126, and 41–52 for Horkheimer's background.

22 Henryk Grossmann, "Die gesellschaftlichen Grundlagen der mechanistischen Philosophie und die Manufaktur," *Zeitschrift für Sozialforschung* 4, no. 2 (1935): 161–231; translated by Gabriella Shalit, ed. Gideon Freudenthal, as "The Social Foundations of Mechanistic Philosophy and Manufacture," *Science in Context* 1 (March 1987): 109–180. The essay is reprinted along with other pertinent writings by Grossmann in Freudenthal and McLaughlin, eds., *Social and Economic Roots*, 103–156.

23 Grossmann, "Social Foundations," esp. 153–156 and 159–170, citation on 154–155.

24 Gideon Freudenthal has pointed out that Grossmann emphasized this aspect of Hessen's paper in a review published in 1938. Freudenthal, "Introductory Note," 105–107. See also Gideon Freudenthal, "Towards a Social History of Newtonian Mechanics: Boris Hessen and Henryk Grossmann Revisited," in *Scientific Knowledge Socialized*, ed. Imre Hronszky, Márta Fehér,

and Balázs Dajka (Dordrecht: Kluwer Academic, 1988), 193–212. Freudenthal and McLaughlin, "Classical Marxist Historiography," esp. 1–20, emphasize the similarity of Grossmann's and Hessen's ideas.

25 For the institute's trajectory from Frankfurt to New York, see especially Wiggershaus, *Frankfurt School*, 127–148; and see Russo, "Henryk Grossmann and Franz Borkenau," 184; Jay, *Dialectical Imagination*, esp. 29–40; and Jay, *Permanent Exiles*, 28–61.

26 In an article on the genesis of the concept of physical law, Zilsel notes that the work of the neo-Kantian philosopher Ernst Cassirer (1874–1945) and especially that of Franz Borkenau is "not quite reliable." Edgar Zilsel, "The Genesis of the Concept of Physical Law," *Philosophical Review* 51 (May 1942): 246 n. 2. For Zilsel's relationship to the Frankfurt School and the influence of the Borkenau/Grossmann controversy, see Hans-Joachim Dahms, "Edgar Zilsels Projekt 'The Social Roots of Science' und seine Beziehungen zur Frankfurter Schule," in *Wien-Berlin-Prag*, 474–500.

27 Edgar Zilsel, "Problems of Empiricism," in *The Development of Rationalism and Empiricism*, vol. 2, no. 8, *International Encyclopedia of United Science*, ed. Otto Neurath (Chicago: University of Chicago Press, 1941), 53–94; Edgar Zilsel, "The Sociological Roots of Science," *American Journal of Sociology* 47 (January 1942): 544–562; his "The Origins of Gilbert's Scientific Method," *Journal of the History of Ideas* 2 (January 1941): 1–32; and "The Genesis of the Concept of Scientific Progress," *Journal of the History of Ideas* 6 (June 1945): 325–349. The above essays are reprinted in Zilsel, *Social Origins of Modern Science*, 171–199, 7–21, and 128–168 (the latter, the original essay, of which the published "Genesis of the Concept of Scientific Progress" was a highly edited version). Zilsel's English-language articles have been collected in a German translation that includes a useful introduction and biographical notes. Edgar Zilsel, *Die sozialen Ursprünge der neuzeitlichen Wissenschaft*, ed. and trans. Wolfgang Krohn, biobibliographical notes by Jörn Behrmann (Frankfurt am Main: Suhrkamp, 1976).

28 See Leonardo Olschki, *Geschichte der neusprachlichen wissenschaftlichen Literatur*, vol. 1: *Die Literatur der Technik und der angewandten Wissenschaften von Mittelalter bis zur Renaissance* (Heidelberg: Winter, 1919), vol. 2: *Bildung und Wissenschaft im Zeitalter der Renaissance in Italien* (Leipzig: Olschki, 1922), and vol. 3: *Galilei und seine Zeit* (Halle: Niemeyer, 1927); and see Cohen, *Scientific Revolution*, esp. 322–326 (citation on 322).

29 Robert K. Merton, *Science, Technology and Society in Seventeenth-Century England* (1938; [Atlantic Highlands], N.J.: Humanities Press, 1978). See also I. Bernard Cohen, "The Publication of *Science, Technology and Society*: Circumstances and Consequences"; Thomas F. Gieryn, "Distancing Science from Religion in Seventeenth-Century England"; and Steven Shapin, "Understanding the Merton Thesis"—all in "Symposium on the Fiftieth Anniversary of *Science, Technology and Society*," *Isis* 79 (December 1988): 571–604. For an insightful discussion of Sarton and the shaping of the history of science in the United States, see Michael Aaron Dennis, "Historiography of

Science: An American Perspective," in *Science in the Twentieth Century,* ed. John Krige and Dominique Pestre (Amsterdam: Harwood Academic, 1997), 1–26.

30 Alexandre Koyré, *From the Closed World to the Infinite Universe* (Baltimore: Johns Hopkins University Press, 1957); and esp. Alexandre Koyré, *Études galiléennes* (Paris: Hermann, 1966). For a recent discussion of Koyré's influence, see esp. Anna-K. Mayer, "Setting Up a Discipline, II: British History of Science and 'The End of Ideology,' 1931–1948," *Studies in the History and Philosophy of Science* 35 (March 2004): 41–72, esp. 61–62.

31 Mayer, "Setting Up a Discipline," shows the explicit anti-Marxian context of these ideas and the ways that they shaped the discipline of the history of science in Britain in the postwar years.

32 A. Rupert Hall, "The Scholar and the Craftsman in the Scientific Revolution," in *Critical Problems in the History of Science,* ed. Marshall Clagett (Madison: University of Wisconsin Press, 1959), 3–23, citations on 3 and 18; and Hall, "Merton Revisited, or Science and Society in the Seventeenth Century," *History of Science: An Annual Review of Literature, Reseach and Teaching,* vol. 2, edited by A.C. Crombie and M. A. Hoskin (Cambridge: W. Heffer and Sons, 1963), 1–16. See also Mayer, "Setting Up a Discipline," 55–65, who discusses Koyré's influence on Hall and shows that Hall's separation of science from the crafts was decisive in his being hired for the position in the history of science offered for the first time at Cambridge. And see Enebakk, "Lilley Revisited,"who discusses Hall's opposition, not only to Merton in the United States, but also to the ideas of the Marxist mathematician and historian of science, Samuel Lilley (1914–1987).

33 Jan Golinski, *Making Natural Knowledge: Constructivism and the History of Science* (Cambridge: Cambridge University Press, 1998), citation on ix. See also Peter Dear, "What Is the History of Science the History Of? Early Modern Roots of the Ideology of Modern Science," *Isis* 96 (September 2005): 390–406.

34 Thomas S. Kuhn, *The Structure of Scientific Revolutions,* 2d ed. (Chicago: University of Chicago Press, 1970). See also Paul Hoyningen-Huene, *Reconstructing Scientific Revolutions: Thomas S. Kuhn's Philosophy of Science,* trans. Alexander T. Levine (Chicago: University of Chicago Press, 1993), 265; and see Golinski, *Making Natural Knowledge,* esp. 13–27. Somewhat ironically, since his views had the effect of disrupting "unity of science," Kuhn's book was published as the last volume of the *International Encyclopedia of Unified Science,* the Vienna Circle series founded by Otto Neurath and others and published by the University of Chicago Press. See esp. George A. Reisch, "Planning Science: Otto Neurath and the *International Encyclopedia of Unified Science,*" *British Journal for the History of Science* 27 (June 1994): 153–175; and Ian Hacking, "The Disunities of the Sciences," and Jordi Cat, Nancy Cartwright, and Hasok Chang, "Otto Neurath: Politics and the Unity of Science," both in *The Disunity of Science: Boundaries, Contexts, and Power,* ed. Peter Galison and David J. Stump (Stanford, Calif.: Stanford University Press, 1996), 37–74 and 347–369, respectively.

35 For a concise summary, see esp. Golinski, *Making Natural Knowledge,* esp. 1–46; and see Massimo Mazzotti, ed., *Knowledge as Social Order: Rethinking*

the *Sociology of Barry Barnes* (Aldershot, Eng.: Ashgate, 2008). For a useful discussion of the French context, see Geof Bowker and Bruno Latour, "A Booming Discipline Short of Discipline: (Social) Studies of Science in France," *Social Studies of Science* 17 (November 1987): 715–748.

36 Steven Shapin and Simon Schaffer, *Leviathan and the Air-Pump* (Princeton, N.J.: Princeton University Press, 1985).

37 Paolo Rossi, *Philosophy, Technology, and the Arts in the Early Modern Era*, trans. Salvator Attanasio, ed. Benjamin Nelson (New York: Harper and Row, 1970; James A. Bennett, "The Mechanics' Philosophy and the Mechanical Philosophy," *History of Science* 24 (March 1986): 1–28; William Eamon, *Science and the Secrets of Nature: Books of Secrets in Medieval and Early Modern Culture* (Princeton, N.J.: Princeton University Press, 1994); Pamela O. Long, *Openness, Secrecy, Authorship: Technical Arts and the Culture of Knowledge from Antiquity to the Renaissance* (Baltimore: Johns Hopkins University Press, 2001); and Pamela H. Smith, *The Body of the Artisan: Art and Experience in the Scientific Revolution* (Chicago: University of Chicago Press, 2004).

38 Paula Findlen, *Possessing Nature: Museums, Collecting, and Scientific Culture in Early Modern Italy* (Berkeley: University of California Press, 1994); William R. Newman and Lawrence M. Principe, *Alchemy Tried in the Fire: Starkey, Boyle, and the Fate of Helmontian Chymistry* (Chicago: University of Chicago Press, 2002); and Tara Nummedal, *Alchemy and Authority in the Holy Roman Empire* (Chicago: University of Chicago Press, 2007).

39 Harold J. Cook, *Matters of Exchange: Commerce, Medicine, and Science in the Dutch Golden Age* (New Haven, Conn.: Yale University Press, 2007); Brian W. Ogilvie, *The Science of Describing: Natural History in Renaissance Europe* (Chicago: University of Chicago Press, 2006); Alix Cooper, *Inventing the Indigenous: Local Knowledge and Natural History in Early Modern Europe* (Cambridge: Cambridge University Press, 2007); and Antonio Barrera-Osorio, *Experiencing Nature: The Spanish American Empire and the Early Scientific Revolution* (Austin: University of Texas Press, 2006); and see Pamela O. Long, "Plants and Animals in History: The Study of Nature in Renaissance and Early Modern Europe," *Historical Studies in the Natural Sciences* 38 (Spring 2008): 313–323.

40 Deborah E. Harkness, *The Jewel House: Elizabethan London and the Scientific Revolution* (New Haven, Conn.: Yale University Press, 2007).

Chapter 2

1 For the mechanical arts, see James A. Bennett, "The Mechanical Arts," in *The Cambridge History of Science*, vol. 3: *Early Modern Science*, ed. Katharine Park and Lorraine Daston (Cambridge: Cambridge University Press, 2006), 673–695; Elspeth Whitney, *Paradise Restored: The Mechanical Arts from Antiquity through the Thirteenth Century*, Transactions of the American Philosophical Society, n.s. 80, pt. 1, 1990. For *ars* and *technē*, see Pamela O. Long, *Openness,*

Secrecy, Authorship: Technical Arts and the Culture of Knowledge from Antiquity to the Renaissance (Baltimore: Johns Hopkins University Press, 2001), 16–45. And see Paul Oskar Kristeller, "The Modern System of the Arts," in Paul Oskar Kristeller, *Renaissance Thought*, II: *Papers on Humanism and the Arts* (New York: Harper and Row, 1965), 163–227.

2 Lorraine Daston, "The Nature of Nature in Early Modern Europe," *Configurations* 6 (Spring 1998): 149–172.

3 Heinrich von Staden, "Physis and Technē in Greek Medicine," and Mark J. Schiefsky, "Art and Nature in Ancient Mechanics," both in *The Artificial and the Natural: An Evolving Polarity*, ed. Bernadette Bensaude-Vincent and William R. Newman (Cambridge, Mass.: MIT Press, 2007), 21–49 and 67–108, respectively.

4 Aristotle, *Physics*, trans. R. P. Hardie and R. K. Gaye in *The Complete Works of Aristotle*, rev. ed., ed. Jonathan Barnes, 2 vols. (Princeton, N.J.: Princeton University Press, 1984), 1:329 (Aristotle, *Physics*, 2.1 [192b9–19]).

5 Ibid., 340 (Aristotle, *Physics*, 191a15–18). It should be underscored that Aristotle did not have a generalized view of nature referring to the natural world and its laws or principles. When he spoke of nature, he usually meant the nature of a thing. See esp. Roger French, *Ancient Natural History: Histories of Nature* (London: Routledge, 1994), esp. 15–18.

6 William R. Newman, *Promethean Ambitions: Alchemy and the Quest to Perfect Nature* (Chicago: University of Chicago Press, 2004), 34–114; and William R. Newman, "Technology and Alchemical Debate in the Late Middle Ages," *Isis* 80 (September 1989): 423–445. For a recent discussion, see Leah DeVun, *Prophecy, Alchemy, and the End of Time: John of Rupescissa in the Late Middle Ages* (New York: Columbia University Press, 2009), esp. 141–148.

7 Newman, *Promethean Ambitions*, 238–289; William R. Newman and Lawrence M. Principe, *Alchemy Tried in the Fire: Starkey, Boyle, and the Fate of Helmontian Chymistry* (Chicago: University of Chicago Press, 2002); and Lawrence M. Principe, *The Aspiring Adept: Robert Boyle and His Alchemical Quest* (Princeton, N.J.: Princeton University Press, 1998).

8 Tara Nummedal, *Alchemy and Authority in the Holy Roman Empire* (Chicago: University of Chicago Press, 2007).

9 Betty Jo Teeter Dobbs, *The Janus Faces of Genius: The Role of Alchemy in Newton's Thought* (Cambridge: Cambridge University Press, 1991).

10 Despite agreement concerning its importance, the precise role of alchemy and many details concerning its characteristics and influence are still matters of contentious debate. See esp. Ursula Klein, "Essay Review: Styles of Experimentation and Alchemical Matter Theory in the Scientific Revolution" *Metascience* 16, no. 2 (2007): 247–256; Brian Vickers, "The 'New Historiography' and the Limits of Alchemy," *Annals of Science* 65 (January 2008): 127–156; and William R. Newman, "Alchemical Atoms or Artisanal 'Building Blocks'? A Response to Klein," and "Brian Vickers on Alchemy and the Occult: A Response," *Perspectives on Science* 17, no. 2 (2009): 212–231 and 17, no. 4 (2009): 482–506, respectively.

11 For a succinct summary of the issue, see David C. Lindberg, "Experiment and Experimental Science," in *The Oxford Dictionary of the Middle Ages*, ed. Robert E. Bjork, 4 vols. (Oxford: Oxford University Press, 2010), 2:604–605; and see Peter Dear, "The Meanings of Experience," in *The Cambridge History of Science*, vol. 3: *Early Modern Science*, ed. Katharine Park and Lorraine Daston (Cambridge: Cambridge University Press, 2008), 106–131. For the discipline of medicine, see Jole Agrimi and Chiara Crisciani, "Per una ricerca su *experimentum-experimenta*: Reflessione epistemologica e tradizione medica (secoli XIII–XV)," in *Presenza del lessico greco e latino nelle lingue contemporanee*, ed. Pietro Janni and Innocenzo Mazzini (Macerata: Università degli Studi di Macerata, 1990), 9–49.

12 Newman, *Promethean Ambitions*, 238–289, citation on 238.

13 Peter Dear, *Discipline and Experience: The Mathematical Way in the Scientific Revolution* (Chicago: University of Chicago Press, 1995). See also Dear, "Meanings of Experience"; and Lorraine Daston and Katharine Park, *Wonders and the Order of Nature, 1150–1750* (New York: Zone Books, 1998).

14 Newman, *Promethean Ambitions*, 238–242. The word "scientist" was invented in the nineteenth century by William Whewell. See Sydney Ross, "Scientist: The Story of a Word," *Annals of Science* 18 (June 1962): 65–85.

15 See esp. G. E. R. Lloyd, *Magic, Reason, and Experience: Studies in the Origin and Development of Greek Science* (Cambridge: Cambridge University Press, 1979); and G. E. R. Lloyd, "Experiment in Early Greek Philosophy and Medicine," in G. E. R Lloyd, *Methods and Problems of Greek Science: Selected Papers* (Cambridge: Cambridge University Press, 1991), 70-99. It should be noted that a group of philosophers insists on the primacy of Aristotle's rationality over his empiricism. See especially Michael Frede, "Aristotle's Rationalism," in *Rationality in Greek Thought*, ed. Michael Frede and Gisela Striker (New York: Oxford University Press, 1996), 157–173; and Joseph Owens, "The Universality of the Sensible in the Aristotelian Noetic," in *Aristotle: The Collected Papers of Joseph Owens*, ed. John R. Catan (Albany: State University of New York Press, 1981), 59–73. I thank the philosopher Jean de Groot for discussing this view of Aristotle's rationalism in the light of her own work in progress, descriptively titled "Aristotle's Empiricism: Experience, Mechanics, and Natural Powers."

16 Dear, *Discipline and Experience*, 24–25. For the complexity of the air-pump and the difficulty others had in building it and replicating Boyle's experiments, see Steven Shapin and Simon Schaffer, *Leviathan and the Air-Pump: Hobbes, Boyle, and the Experimental Life* (Princeton, N.J.: Princeton University Press, 1985), esp. 225–282.

17 Pamela H. Smith, *The Body of the Artisan: Art and Experience in the Scientific Revolution* (Chicago: University of Chicago Press, 2004), esp. 100–106 (Palissy). See also Pamela H. Smith and Tonny Beentjes, "Nature and Art, Making and Knowing: Reconstructing Sixteenth-Century Life-Casting Techniques," *Renaissance Quarterly* 63 (Spring 2010): 128–179

18 See esp. Daston and Park, *Wonders and the Order of Nature*, 255–301; Paula Findlen, *Possessing Nature: Museums, Collecting, and Scientific Culture in Early*

Modern Italy (Berkeley: University of California Press, 1994); Martin Kemp, "Wrought by No Artist's Hand: The Natural, the Artificial, the Exotic, and the Scientific in Some Artifacts from the Renaissance," in *Reframing the Renaissance: Visual Culture in Europe and Latin America, 1450-1650*, ed. Claire Farago (New Haven, Conn.: Yale University Press, 1995), 177–196; Pamela O. Long, "Objects of Art/Objects of Nature: Visual Representation and the Representation of Nature," in *Merchants and Marvels: Commerce, Science, and Art in Early Modern Europe*, ed. Pamela H. Smith and Paula Findlen (New York: Routledge, 2002), 63–82. For Arcimboldo, see esp. Thomas DaCosta Kaufmann, *Arcimboldo: Visual Jokes, Natural History, and Still-Life Painting* (Chicago: University of Chicago Press, 2009), esp. 43–68 and 115–166, who has much to say concerning the art/nature relationship in Arcimboldo's work and surrounding culture..

19 Lisa Jardine, *Worldly Goods: A New History of the Renaissance* (New York: Doubleday, 1996).

20 For the maker's knowledge tradition, see esp. Antonio Pérez-Ramos, *Francis Bacon's Idea of Science and the Maker's Knowledge Tradition* (Oxford: Clarendon, 1988); and for the view of the pre-Baconian development of such a tradition, see, for example, Deborah E. Harkness, *The Jewel House: Elizabethan London and the Scientific Revolution* (New Haven, Conn.: Yale University Press, 2007); and Long, *Openness, Secrecy*, esp. 102–142 and 175–250.

21 The attribution to Francesco Colonna is supported by an acrostic formed by the elaborately decorated initial letters of the thirty-eight chapters of the book which reads: poliam frater franciscus columna peramavit ("Brother Francesco Colonna greatly loved Polia"). Arguments for various other authors, including Leon Battista Alberti, have not been convincing. My discussion is based on the following edition and commentary: Francesco Colonna, Hypnerotomachia Poliphili, ed. Giovanni Pozzi and Lucia A. Ciapponi, 2 vols. (Padua: Editrice Antenore, 1964). Translations are my own. For Francesco Colonna's life and a discussion of the work, see Maria T. Casella and Giovanni Pozzi, Francesco Colonna: Biografia e opera, 2 vols. (Padua: Editrice Antenore, 1959). For the (unconvincing) claim that the author was Alberti, see Liane Lefaivre, Leon Battista Alberti's Hypnerotomachia Poliphili: Re-Cognizing the Architectural Body in the Early Italian Renaissance (Cambridge, Mass.: MIT Press, 1997). See also Helen Barolini, Aldus and His Dream Book (New York: Italica Press, 1992); and Alberto Pérez-Gómez, "The Hypnerotomachia Poliphili by Francesco Colonna: The Erotic Nature of Architectural Meaning," in Paper Palaces: The Rise of the Renaissance Architectural Treatise, ed. Vaughan Hart with Peter Hicks (New Haven, Conn.: Yale University Press, 1998), 86–104; For an English translation see Francesco Colonna, Hypnerotomachia Poliphili: The Strife of Love in a Dream, trans. Joscelyn Godwin (New York: Thames and Hudson, 1999).

22 Colonna, Hypnerotomachia Poliphili, ed. Pozzi and Ciapponi, 1:19–28, citation on 28, "intestine, nervi et ossa, vene, musculi et pulpamento."

23 Ibid., 1: 34: "vedevsi quasi il tremulare degli sui pulpamenti, et più vivo che fincto"; and 53–95.

24 Ibid., 1: 112–123, citation on 116: "omni pianta era di purgatissimo vitro, egregiamente oltra quello che se pole imaginare et credere, intopiati buxi cum gli stirpi d'oro."

25 For Francesco, see esp. Richard J. Betts, "On the Chronology of Francesco di Giorgio's Treatises: New Evidence from an Unpublished Manuscript," *Journal of the Society of Architectural Historians* 36 (March 1977): 3–14; Francesco Paolo Fiore and Manfredo Tafuri, eds., *Francesco di Giorgio architetto* (Milan: Electa, 1993); Gustina Scaglia, *Francesco di Giorgio: Checklist and History of Manuscripts and Drawings in Autographs and Copies from ca. 1470 to 1687 and Renewed Copies (1764–1839)* (Bethlehem, Penn.: Lehigh University Press; and Cranbury, N.J.: Associated University Presses, 1992); and Ralph Toledano, *Francesco di Giorgio Martini: Pittore e scultor.* (Milan: Electa, 1987).

26 Francesco di Giorgio Martini, Trattati di architettura ingegneria e arte militare, 2 vols., ed. Corrado Maltese, transcription by Livia Maltese Degrassi (Milan: Edizioni Il Polifilo, 1967). See also Massimo Mussini, Il Trattato di Francesco di Giorgio Martini e Leonardo: Il Codice Estense restituito (Parma: Università di Parma, 1991), 82–88, 108–109, 121–124, nn. 75–79; and Massimo Mussini, "Un frammento del Trattato di Francesco di Giorgio Martini nell'archivio di G. Venturi alla Biblioteca Municipale di Reggio Emilia," in Prima di Leonardo: Cultura delle macchine a Siena nel Rinascimento, ed. Paolo Galluzzi (Milan: Electa, 1991), 81–92.

27 Francesco di Giorgio, Trattati di architettura ingegneria e arte militare, vol. 1, which includes a facsimile of the sheets containing drawings of T. For a brief but masterly description of the manuscripts and the scholarship surrounding their origins and dating, see Massimo Mussini, "La trattatistica di Francesco di Giorgio: un problema critico aperto," in Francesco di Giorgio architetto, ed. Fiore and Tafuri, 358–379. See also Scaglia, Francesco di Giorgio: Checklist, 154–159 (no. 62, for Manuscript L) and 189–191 (no. 80, for Manuscript T), although it should be noted that Scaglia's claim that the two manuscripts were created at the monastery of Monte Oliveto Maggiore is without foundation in evidence and has been cogently disputed by Mussini. A further discussion of the complex issue of the chronology of Francesco's writings and a summary of the scholarship is in Marco Biffi, "Introduzione," in Francesco di Giorgio Martini, *La traduzione del "De architectura" di Vitruvio dal ms. II.I.141 della Biblioteca Nazionale Centrale di Firenze* (Pisa: Scuola Normale Superiore, 2002), XI–CXVII, esp. XXX–LXVII.

28 My discussion of Francesco di Giorgio and Leonardo is derived from Pamela O. Long, "Picturing the Machine: Francesco di Giorgio and Leonardo da Vinci in the 1490s," in *Picturing Machines, 1400–1700*, ed. Wolfgang Lefèvre (Cambridge, Mass.: MIT Press, 2004), 117–141.

29 Francesco di Giorgio, Trattati di architettura ingegneria e arte militare , vol. 1, fols. 33v–40r of facsimile pages, tav. 62–75. See also Vittorio Marchis, "Nuove dimensioni per l'energia: le macchine di Francesco di Giorgio," in Prima di Leonardo, ed. Galluzzi, 113–120.

30 Francesco di Giorgio, *Trattati di architettura ingegneria e arte militare*, vol. 1, xi–xlviii. See also Mussini "La trattatistica di Francesco di Giorgio," 358–359.

31 Francesco di Giorgio, Trattati di architettura ingegneria e arte militare, 2:500–501; and fol. 95 (tav. 325). The Magliabechiano manuscript contains several other works by Francesco as well—the *Trattato II* is on fols. 1–102; Francesco's translation of Vitruvius on fols. 103–192; and a collection of drawings of military machines and fortification designs is on fols. 193–244v.

32 Francesco di Giorgio, *Trattati di architettura ingegneria e arte militare*, 2:492–504, citation on 492, "Si ancora di alquanti pistrini metterò la figura acciò che, per quelli, delli altri simili da li lettori possino essere trovati."

33 Ibid., 2:500–501 and fol. 95 (tav. 325).

34 Ibid.

35 See Paolo Galluzzi, "The Career of a Technologist," and Salvatore Di Pasquale, "Leonardo, Brunelleschi and the Machinery of the Construction Site," both In *Leonardo da Vinci: Engineer and Architect*, ed. Paolo Galluzzi (Montreal: Montreal Museum of Fine Arts, 1987), 41–109, esp. 48–63, for Leonardo's early career in Florence, and 163–181, for Leonardo's study of Brunelleschi's construction machinery. An excellent general introduction to Leonardo's work as a whole and to some of the vast scholarship is Martin Kemp, *Leonardo da Vinci: The Marvellous Works of Nature and Man*, rev. ed. (Oxford: Oxford University Press, 2006).

36 Leonardo da Vinci, *The Madrid Codices*, 5 vols., trans. Ladislao Reti (New York: McGraw-Hill, 1974), 3:11–21, for the history of the volumes. And see Robert Zwijnenberg, *The Writings and Drawings of Leonardo da Vinci: Order and Chaos in Early Modern Thought*, trans. Caroline A. van Eck (Cambridge: Cambridge University Press, 1999), esp. 83–111.

37 Leonardo, Madrid Codices, trans. Reti, 1:15v and 4:39–40.

38 Ibid.

39 Ibid., 1:15v and 4:41–42.

40 Ibid.

41 See esp. Cesare S. Maffioli, *La via delle acque (1500–1700): Appropriazione delle arti e trasformazione delle matematiche* (Florence: Leo S. Olschki, 2010), for the transformation of the mathematical arts that led to the mechanics of Galileo and his successors.

42 Sebastiano Serlio, *Regole generali di architectura sopra le cinque maniere de gliedifici, cioe, thoscano, dorico, ionico, corinthio, et composito, con gliessempi dell'antiquita, che per la magior parte concordano con la dottrina di Vitruvio* (Venice: F. Marcolini da Forli, 1537); and Sebastiano Serlio, *On Architecture*, vol. 1: *Books I–V of "Tutte l'opere d'architettura et prospetiva*," and vol. 2: *Books VI and VII of "Tutte l'opere d'architettura et prospetiva" with "Castrametation of the Romans" and "The Extraordinary Book of Doors" by Sebastiano Serlio*, trans. with introduction and commentary by Vaughan Hart and Peter Hicks (New Haven, Conn.: Yale University Press, 1996, 2001), an English translation with extensive commentary and bibliography. For Serlio's life and the complex publishing history of his writings, the fundamental work remains William Bell Dinsmoor,

"The Literary Remains of Sebastiano Serlio," *Art Bulletin* 24 (March 1942): 55–91. More recent scholarship includes Vaughan Hart and Peter Hicks, "On Sebastiano Serlio: Decorum and the Art of Architectural Invention," in *Paper Palaces*, ed. Hart with Hicks, 140–157; Myra Nan Rosenfeld, "From Bologna to Venice and Paris: The Evolution and Publication of Sebastiano Serlio's Books I and II, *On Geometry* and *On Perspective*, for Architects," in *The Treatise on Perspective: Published and Unpublished*, ed. Lyle Massey (Washington, D.C.: National Gallery of Art, 2003), 280–321; and Alina A. Payne, *The Architectural Treatise in the Italian Renaissance: Architectural Invention, Ornament, and Literary Culture* (Cambridge: Cambridge University Press, 1999), 113–143.

43 Andreas Vesalius, *De humani corporis fabrica libri septem* (Basil: [Ex officina I. Oporini, 1543]). For English translations, see Andreas Vesalius, *On the Fabric of the Human Body: A Translation of "De humani corporis fabrica libri septem,"* 7 vols., trans. William Frank Richardson and John Burd Carman (San Francisco: Norman, 1998–2009); and Andreas Vesalius, *De humani corporis fabrica*, trans. Daniel H. Garrison and Malcolm Hast (in progress), http://vesalius.northwestern.edu (accessed December 2010). For Vesalius, see esp. Andrea Carlino, *Books of the Body: Anatomical Ritual and Renaissance Learning*, trans. John Tedeschi and Anne C. Tedeschi (Chicago: University of Chicago Press, 1999), esp. 39–53 and 201–213; Andrew Cunningham, *The Anatomical Renaissance: The Resurrection of Anatomical Projects of the Ancients* (Aldershot, Eng.: Scolar Press, 1997), 88–142; C. D. O'Malley, *Andreas Vesalius of Brussels, 1514–1564* (Berkeley: University of California Press, 1964); Vivian Nutton, "Introduction," Vesalius, *De humanis corporis fabrica*, ed. Garrison and Hast, http://vesalius.northwestern.edu; Katharine Park, *Secrets of Women: Gender, Generation, and the Origins of Human Dissection* (New York: Zone Books, 2006), 207–259; Nancy G. Siraisi, "Vesalius and Human Diversity in *De humani corporis fabrica*," *Journal of the Warburg and Courtauld Institutes* 57 (1994): 60–88; Nancy G. Siraisi, "Vesalius and the Reading of Galen's Teleology," *Renaissance Quarterly* 50 (Spring 1997): 1–37; and Andrew Wear, "Medicine in Early Modern Europe, 1500–1700," in *The Western Medical Tradition, 800 BC to AD 1800*, ed. Lawrence C. Conrad, Michael Neve, Vivian Nutton, Roy Porter, and Andrew Wear (Cambridge: Cambridge University Press, 1995), 207–361, esp. 273–279.

44 Serlio, *On Architecture*, ed. Hart and Hicks, 1:xix.

45 Andreas Vesalius, *Tabulae anatomicae sex* (Venice: B. Vitalis, 1538). For Calcar, see Bert W. Meijer, "Calcar, Jan Steven [Johannes Stephanus] van," *The Dictionary of Art*, ed. Jane Turner, 34 vols. (New York: Grove Dictionaries, 1996), 5: 415-416. For the complex issue of the identity of the illustrators, see esp. Martin Kemp, "A Drawing for the *Fabrica*; and Some Thoughts upon the Vesalius Muscle-Men," *Medical History* 14 (July 1970): 277–288; Michelangelo Muraro, "Tiziano e le anatomie del Vesalio," in *Tiziano e Venezia: Convegno Internazionale di Studi, Venezia, 1976* (Vicenza: Neri Pozza, 1980), 307–316; David Rosand and Michelangelo Muraro, *Titian and the Venetian Woodcut* (Washington, D.C.: International Exhibitions Foundation, 1976–1977), 211–

235; and Patricia Simons and Monique Kornell, "Annibal Caro's After-Dinner Speech (1536) and the Question of Titian as Vesalius's Illustrator," *Renaissance Quarterly* 61 (Winter 2008): 1069–1097.

46 The association with the Belvedere statue was first made by H. W. Janson, "Titian's Laocoon Caricature and the Vesalian-Galenist Controversy," *Art Bulletin* 28 (March 1946): 51. For a detailed discussion, see esp. Glenn Harcourt, "Andreas Vesalius and the Anatomy of Antique Sculpture," *Representations* 17 (Winter 1987): 28–61. For Polycletus and Vesalius, see esp. Jackie Pigeaud, "Formes et normes dans le 'De fabrica' de Vésale," in *Le corps à la Renaissance, Actes du XXXe Colloque de Tours, 1987,* ed. Jean Céard, Marie Madeleine Fontaine, and Jean-Claude Margolin (Paris: Aux Amateurs de Livres, 1990), 399–421; and Catrien Santing, "Andreas Vesalius's *De Fabrica corporis humana,* Depiction of the Human Model in Word and Image," in *Body and Embodiment in Netherlandish Art,* Netherlands Yearbook for History of Art, 2007–2008, vol. 58, ed. Ann-Sophie Lehmann and Herman Roodenburg (Zwolle: Waanders, 2008), 59–85. For a broader context, see Martin Kemp, "'The Mark of Truth': Looking and Learning in Some Anatomical Illustrations from the Renaissance and Eighteenth Century," in *Medicine and the Five Senses,* ed. W. F. Bynum and Roy Porter (Cambridge: Cambridge University Press, 1993), 85–121.

47 Serlio, *On Architecture,* trans. Hart and Hicks, 1:266–269.

48 Ibid., 1:286–287.

49 Vesalius, *On the Fabric,* bk. 1: *Bones and Cartilages,* trans. Richardson and Carman, "To King Charles V," xlvii–xlix.

50 For an insightful discussion of this image and its meaning in relationship to the famous frontispiece in which Vesalius is depicted dissecting the corpse of a woman in a crowded anatomy hall, see Park, *Secrets of Women,* esp. 249–255. For Vesalius's interest in both natural philosophy, or *scientia,* and hands-on observation and investigation through dissection, or *ars,* see esp. Siraisi, "Vesalius and Human Diversity," 65–67.

51 Vesalius, *On the Fabric,* bk. 1: *Bones and Cartilages,* trans. Richardson and Carman, "To King Charles V," liv–lv.

52 Ibid., lvi. The reference is to Galen, *Procedures,* 2.1.

53 Ibid., 1:1 and 8.

54 Ibid., 1:4.

Chapter 3

1 Vitruvius, *On Architecture,* ed. and trans. Frank Granger, 2 vols., Loeb (Cambridge, Mass.: Harvard University Press, 1931–1934). For a more recent English translation of the text and commentary with helpful drawings, see Vitruvius, *Ten Books on Architecture,* ed. Ingrid D. Rowland and Thomas Noble Howe (Cambridge: Cambridge University Press, 1999), 3–5, for the dating of the treatise. I have used Rowland's translation. A recent study and interpretation that places the work within the context of the Roman empire is

Indra Kagis McEwen, *Vitruvius: Writing the Body of Architecture* (Cambridge, Mass.: MIT Press, 2003). An erudite synthetic account of Vitruvian influence and architectural writing in the Renaissance is Alina A. Payne, *The Architectural Treatise in the Italian Renaissance: Architectural Invention, Ornament, and Literary Culture* (Cambridge: Cambridge University Press, 1999).

2 Vitruvius, *On Architecture*, ed. and trans. Granger, 1:4–5 (*De arch*. 1. pref. 2–3); and see Vitruvius, *Ten Books*, ed. Rowland and Howe, 21 and 135.

3 Vitruvius, *On Architecture*, ed. and trans. Granger, 1:151–247 (*De arch*. bks. 3 and 4—temples), 1:249–317 (*De arch*. bk. 5—public buildings), 2:1–59 (*De arch*. bk. 6—private buildings), 1:34–67 (*De arch*. 1.4.–1.6—siting and layout of cities), 1:88–149 (*De arch*. 2.3–10—building materials), 2:80–87 (*De arch*. 7.1—flooring), 2:88–97 (*De arch*. 7.3—ceilings), 2:96–129 (*De arch*. 7.4–14—painting and colors), and 2:86–101 (*De arch*. 7.2–7.4—stucco).

4 Ibid, 2:131–193 (*De arch*. bk. 8); and see Louis Callebat, ed. and trans., *Vitruve De L'architecture, Livre VIII* (Paris: Société D'Édition "Les Belle Lettres," 1973).

5 Vitruvius, *On Architecture*, ed. and trans. Granger , 2:195–267 (*De arch*. bk. 9); and see Jean Soubiran, ed. and trans., *Vitruve De L'architecture, Livre IX* (Paris: Société D'Édition "Les Belle Lettres," 1969).

6 Vitruvius, *On Architecture*, ed. and trans. Granger, 2:269–369 (*De arch*. bk. 10); and see Louis Callebat and Philippe Fleury, eds. and trans., *Vitruve De L'architecture, Livre X* (Paris: Société D'Édition "Les Belle Lettres," 1986).

7 Vitruvius, *On Architecture*, ed. and trans. Granger, 2:6–7 (*De arch*. 6. pref. 5); Vitruvius, *Ten Books on Architecture*, ed. Rowland and Howe, 75–76, 249.

8 Vitruvius, *On Architecture*, ed. and trans. Granger, 1:6–7 (*De arch*. 1.1.1–2.); Vitruvius, *Ten Books on Architecture*, ed. Rowland and Howe, 21 (translation slightly altered).

9 Vitruvius, *On Architecture*, ed. and trans. Granger, 1:8–25 (*De arch*. 1.1.4–18); Vitruvius, *Ten Books on Architecture*, ed. Rowland and Howe, 22–24, 135–143.

10 Vitruvius, *On Architecture*, ed. and trans. Granger, 2:4–7 (*De arch*. 6. pref. 4.); Vitruvius, *Ten Books on Architecture*, ed. Rowland and Howe, 75, 249.

11 Vitruvius, *On Architecture*, ed. and trans. Granger, 1:76–85 (*De arch*. 2.1.1–7); Vitruvius, *Ten Books on Architecture*, ed. Rowland and Howe, 34–35, 175. For the anthropological tradition in antiquity see Thomas Cole, *Democritus and the Sources of Greek Anthropology* (Chapel Hill, N.C.: American Philological Association, 1967).

12 Eugene Dwyer, Peter Kidson, and Pier Nicola Pagliara, "Vitruvius," in *The Dictionary of Art*, 34 vols., ed. Jane Turner (New York: Grove Dictionaries, 1996), 32:632–643. For Faventinus, see Hugh Plommer, *Vitruvius and Later Roman Building Manuals* (London: Cambridge University Press, 1973). For the medieval manuscript tradition, see especially Carol H. Krinsky, "Seventy-Eight Vitruvius Manuscripts," *Journal of the Warburg and Courtauld Institutes* 30 (1967): 36–70.

13 Pamela O. Long, "The Contribution of Architectural Writers to a 'Scientific' Outlook in the Fifteenth and Sixteenth Centuries," *Journal of Medieval and Renaissance Studies* 15 (Fall 1985): 265–298.

14 For a popular but detailed account, see Paul Robert Walker, *The Feud That Sparked the Renaissance: How Brunelleschi and Ghiberti Changed the Art World* (New York: HarperCollins, 2002).

15 Antonio di Tuccio Manetti, *The Life of Brunelleschi*, introduction, notes, and critical text by Howard Saalman and trans. Catherine Enggass (University Park: Pennsylvania State University Press, 1970), 36–41. The dome of the Florentine Cathedral has engendered much debate concerning Brunelleschi's methods. See esp. Howard Saalman, *Filippo Brunelleschi: The Cupola of Santa Maria del Fiore* (London: Zwemmer, 1980); and Barry Jones, Andrea Sereni, and Massimo Ricci, "Building Brunelleschi's Dome: A Practical Methodology Verified by Experiment," *Journal of the Society of Architectural Historians* 69 (March 2010): 39–61. A good popular account is Ross King, *Brunelleschi's Dome: How a Renaissance Genius Reinvented Architecture* (New York: Penguin Books, 2001).

16 Manetti, *Life of Brunelleschi*, 50–53, citation on 50–51.

17 Ibid., 52–53.

18 Ibid., 54–55.

19 Howard Saalman, introduction to Manetti, *Life of Brunelleschi*, 26, 30.

20 For Brunelleschi's machines, including his patented ship, see esp. Christine Smith, *Architecture in the Culture of Early Humanism: Ethics, Aesthetics, and Eloquence, 1400–1470* (Oxford: Oxford University Press, 1992), 19–39; Frank D. Prager and Gustina Scaglia, *Brunelleschi: Studies of His Technology and Inventions* (Cambridge, Mass.: MIT Press, 1970); and for the long-lived influence of Brunelleschi's machines in Florence, Salvatore di Pasquale, "Leonardo, Brunelleschi, and the Machinery of the Construction Site," in *Leonardo da Vinci: Engineer and Architect*, ed. Paolo Galluzzi (Montreal: Montreal Museum of Fine Arts, 1987), 163–181.

21 See Martin Kemp, introduction to Leon Battista Alberti, *On Painting*, trans. Cecil Grayson (London: Penguin Books, 1991), 1–29, on which this summary is based. See also Stefano Borsi, *Leon Battista Alberi e Roma* (Florence: Edizioni Polistampa, 2003; Stefano Borsi, *Leon Battista Alberti e l'antichità Romana* (Florence: Edizioni Polistampa, 2004); Joan Gadol, *Leon Battista Alberti: Universal Man of the Early Renaissance* (Chicago: University of Chicago Press, 1969), a still useful account especially for Alberti's technical achievements; Anthony Grafton, *Leon Battista Alberti: Master Builder of the Italian Renaissance* (New York: Hill and Wang, 2000); and Paul Davies and David Hemsoll, "Alberti, Leon Battista," *Dictionary of Art*, ed. Turner, 1:555–569, for a summary of Alberti's treatises on painting, sculpture, and architecture, and a succinct treatment of his architectural works. See also Alessandro Gambuti, "Nuove ricerche sugli Elementa Picturae," and Luigi Vagnetti, "Considerazioni sui Ludi Matematici," both in *Studi e Documenti di Architettura*, no. 1 (December 1972): 131–172 and 173–259, respectively. And see Mario Carpo and Francesco Furlan, eds., *Leon Battista Alberti's Delineation of the City of Rome (Descriptio urbis Romae)*, critical edition by Jean-Yves Boriaud and Francesco Furlan, trans. Peter Hicks (Tempe: Arizona Center for Medieval and Renaissance Studies, 2007).

22 See Carroll William Westfall, *In This Most Perfect Paradise: Alberti, Nicholas V, and the Invention of Conscious Urban Planning in Rome, 1447–55* (University Park: Pennsylvania State University Press, 1974).

23 Smith, *Architecture in the Culture of Early Humanism*, 98–132.

24 Leon Battista Alberti, *On Painting*, introduction and notes by Martin Kemp, trans. Cecil Grayson (London: Penguin Books, 1991); and Rocco Sinisgalli, *Il nuovo "De Pictore" di Leon Battista Alberti / The New "De pictore" of Leon Battista Alberti* (Rome: Edizioni Kappa, 2006).

25 "From Angelo Poliziano to his Patron, Lorenzo de' Medici, Greetings," in Leon Battista Alberti, *On the Art of Building in Ten Books*, trans. Josph Rykwert, Neil Leach, and Robert Tavernor (Cambridge, Mass.: MIT Press, 1988), 1. For the Latin text with Italian translation, see Leon Battista Alberti, *L'architettura [De re Aedificatoria]*, 2 vols., ed. and trans. Giovanni Orlandi, introduction and notes by Paolo Portoghesi (Milan: Edizioni il Polifilo,1966).

26 Alberti, *On the Art of Building*, 3.

27 Ibid., 3–4.

28 Ibid., 33–60 (bk. 2), 61–91 (bk. 3), 92–116 (bk. 4, public works), 117–153 (bk. 5, private works), 164–175 (tools and machines), 344–349 (roads, highways, canals).

29 See Walker, *Feud That Sparked the Renaissance.*

30 For a succinct discussion, see Manfred Wundram, "(1) Lorenzo (di Cione) Ghiberti," *Dictionary of Art*, ed. Turner, 12:536–545; and for the third set of doors, Gary M. Radke, ed. *The Gates of Paradise: Lorenzo Ghiberti's Renaissance Masterpiece* (Atlanta: High Art Museum; New Haven, Conn.: Yale University Press, 2007). The relief portrait of Brunelleschi was made by Lazzaro Cavalcanti known as Buggiano circa 1446.

31 Lorenzo Ghiberti, *I commentarii (Biblioteca Nazionale Centrale di Firenze, II, I, 333)*, ed. Lorenzo Bartoli (Florence: Giunti Gruppo Editoriale, 1998). And see Wundram, "(1) Lorenzo (di Cione) Ghiberti," 12:543; and Richard Krautheimer, in collaboration with Trude Krautheimer-Hess, *Lorenzo Ghiberti*, 3d ed. (Princeton, N.J.: Princeton University Press, 1982), 306–314.

32 Ghiberti, *I commentarii*, ed. Bartoli, 46 (II.1.), "Conviene che 'llo scultore, etiamdio el picture, sia amaestrato in tutte queste arti liberali."

33 Ibid., 46 (II.2), "L'isculura et pictura è scientia di più discipline et di varii amaestramenti ornata, la quale di tutte l'altre arti è somma invention. È fabricata con certa meditatione, la quale si compie per materia et ragionamenti"; 47 (II.3), "Et così gli scultori et pictori gli quali sanza lettere aviano conteso come se colle mani avessino exercitato, non poterono compiere né finire come se avessono avuta l'autorità per le fatiche; et quelli i quali per ragionamenti et con lettere sole si veggono conquisi ànno l'ombra, ma non la cosa." For a useful discussion, see Krautheimer with Krautheimer-Hess, *Lorenzo Ghiberti*, 306–309.

34 Filarete, *Filarete's Treatise on Architecture*, 2 vols., ed. and trans. John R. Spencer (New Haven, Conn.: Yale University Press, 1965); and Antonio Averlino detto il Filarete, *Trattato di Architectura*, 2 vols., ed. Anna Maria Finoli and

Liliana Grassi (Milan: Edizioni il Polifilo, 1972), LXXXIX (for his attempted theft of the head of St. John the Baptist). For a recent group of substantial studies, see the special issue of *Arte Lombarda*—Berthold Hub, ed., *Architettura e umanismo: Nuovi studi su Filarete, Arte Lombarda,* n.s. 155, 1 (2009).

35 Although the original manuscript is lost, several copies survive, of which, scholars agree, the best is Florence, Biblioteca Nazionale Centrale, Magliabechianus II, I, 140.

36 See Maria Beltramini, "Francesco Filelfo e il Filarete: Nuovi contributi alla storia dell'amicizia fra il letterato e l'architetto nella Milano sforzesca," *Annali della Scuola Normale Superiore di Pisa, Classe di Lettere e Filosofia,* ser. 4, no. 1 (1996): 119–125; S. Lang, "Sforzinda, Filarete and Filelfo," *Journal of the Warburg and Courtauld Institutes* 35 (1972): 391–397; and John Onians, "Alberti and Filarete: A Study in Their Sources," *Journal of the Warburg and Courtauld Institutes* 34 (1971): 96–114, who documents both the friendship and the relevant Greek sources for Filarete's ideal city. A recent astute essay on Filelfo that provides essential further bibliography and stresses the growing esteem for practical knowledge in Italian princely courts is Margaret Meserve, "Nestor Denied: Francesco Filelfo's Advice to Princes on the Crusade against the Turks," in *Expertise: Practical Knowledge and the Early Modern State,* ed. Eric H. Ash, *Osiris,* 2d ser. 25 (2010): 47–65.

37 Filarete, *Treatise on Architecture,* trans. Spencer, 1:6; Filarete, *Trattato,* ed. Finoli and Grassi, 1:13, "sanno mettere una pietra in calcina e imbrattarla di malta, pare loro essere ottimi maestri d'architettura"; "nè misure, nè proporzioni."

38 Filarete, *Treatise on Architecture,* trans. Spencer, 1:16–17; Filarete, *Trattato,* ed. Finoli and Grassi, 1:41–44, "la volentà e le misure."

39 Filarete, *Treatise on Architecture,* trans. Spencer, 1:198; Filarete, *Trattato,* ed. Finoli and Grassi, 2:428, "di più esercizii intendere, e anche coll'opera della mano dimostrarle, con ragioni di misure e di proporzioni e di qualità, e conveniente"; "di sua mano, non saprà mai mostrare, nè dare a 'ntendere cosa che stia bene."

40 Filarete, *Treatise on Architecture,* trans. Spencer, 1:228; Filarete, *Trattato,* ed. Finoli and Grassi, 2:495, "Benchè non abbino tanta dignità"; "qui stia di più esercizii di mano e anche di persona"; "cioè di begli vasi."

41 Filarete, *Treatise on Architecture,* trans. Spencer, 1:231; Filarete, *Trattato,* ed. Finoli and Grassi, 2:500–501, "Questa sarà una cosa che sempre durerà e una cosa che mai non fu fatta"; "non è se none in lettere questa comodità"; "gli altri esercizii sono di necessità e degni, che ne viene buono maestro, et ancora gl'ingegni non sono tutti a una cosa iguali. Si che è si vuole che ogni ingegno si possa esercitare." For a detailed recent study of the social and educational ideas in the treatise and their relationship to Milanese society and culture, see Hubertus Günther, "Society in Filarete's *Libro architettonico* between Realism, Ideal, Science Fiction and Utopia," in *Architettura e umanismo,* ed. Hub, 56–80.

42 Filarete, *Treatise on Architecture,* trans. Spencer, 1:245–255; Filarete, *Trattato,* ed. Finoli and Grassi, 2:531–562, "E in questo luogo di quante arti o esercizii che fare si possino in questo luogo sono tutte" (545); "erano onorati della loro

acquistata virtù" (547); "erano giudicati buoni maestri, se fusse stato giovane e fusse che in questo luogo avesse imparato, come i dottori s'adottoravano" (552).

43 Marco Biffi, "Introduzione," in Francesco di Giorgio Martini, *La traduzione del 'De architectura' di Vitruvio dal ms. II.I.141 della Biblioteca Nazionale Centrale di Firenze* ([Pisa: Scuola Normale Superiore, 2002), XI–CXVII, esp. LXIV (prepared for own use), LXXXVIII–LXXXIX (language), and CV (improvement over the years). See also Massimo Mussini, *Francesco di Giorgio e Vitruvio: Le traduzioni del "De Architectura" nei codici Zichy, Spencer 129 e Magliabechiano II.I.141*, 2 vols. (Mantua: Fondazione Centro Studi L. B. Alberti, 2002). For Francesco's architecture, see Francesco Paolo Fiore and Manfredo Tafuri, eds., *Francesco di Giorgio, architetto* (Milan: Electa, 1993).

44 Francesco di Giorgio, *Trattati di architettura ingegneria e arte militare*, 2 vols., ed. Corrado Maltese, transcription by Livia Maltese Degrassi (Milan: Edizioni Il Polifilo, 1967), 1:36–39; "l'architettura è solo una sottile immaginazione concetta in nella mente la quale in nell'opra si manifesta" (36); "d'ogni e ciascuna cosa non si può la ragione assegnare, perchè lo ingegno consiste più in nella mente e in nello intelletto dell'architettore che in iscrittura o disegno, e molte cose accade in fatto le quali l'architetto overo opratore mai pensò"; "arroganti e presentuosi"; nelli errori fondati sono"; per forza della lingnia [*sic*, the editor argues persuasively for lingua rather than linia (linea)]"; "el mondo hanno corrotto" (36).

45 Giovanni Sulpicius, ed., *L. Vitruvii Pollionis ad Cesarem Augustum De Architectura Liber Primus (—Decimus)* (Rome: [Giorgio Herolt or Eucarius Silber], [1486–92?]. Leon Battista Alberti *De re aedificatoria* (Florence: Nicolaus Larentii, [1486]).

46 For Pomponio Leto's academy, see John F. D'Amico, *Renaissance Humanism in Papal Rome: Humanists and Churchmen on the Eve of the Reformation* (Baltimore: Johns Hopkins University Press, 1983), 91–102; Eugenio Garin, "La letteratura degli umanisti," in *Storia della Letteratura Italiana: Il Quattrocento e l'Ariosto*, rev. ed., ed. Lucio Felici (Milan: Garzanti, 1988), 3:7–368, esp. 144–160; and Ingrid D. Rowland, *The Culture of the High Renaissance: Ancients and Moderns in Sixteenth-Century Rome* (Cambridge: Cambridge University Press, 1998), 10–25.

47 On Sulpizio's edition, see Dwyer, Kidson, and Pagliara, "Vitruvius," in *Dictionary of Art*, ed. Turner, 32:639–640; Luigi Vagnetti and Laura Marcucci, "Per una coscienza vitruviana. Registo cronologico e critico delle edizioni, delle traduzioni e delle ricerche più importanti sul trattato Latino *De architectura Libri X* di Marco Vitruvio Pollione," *Studi e Documenti di Architettura*, no. 8 (September 1978): 29–30; and Laura Marcucci, "Giovanni Sulpicio e la prima edizione del *De architectura* di Vitruvio," *Studi e Documenti di Architettura*, no. 8 (September 1978): 185–195. On the theater production, see Ingrid D. Rowland, introduction to Vitruvius, *Ten Books on Architecture: The Corsini Incunabulum with the Annotations and Autograph Drawings of Giovanni Battista da Sangallo*, ed. Ingrid D. Rowland (Rome: Edizioni dell'Elefante, 2003), 1–31; the performance

was saved by the "plaintive royal laments in improvised verse" made during the set reconstruction by a sixteen-year-old student in flowing robes playing the queen—Tomasso Inghirami (1470–1516) (7).

48 Sulpicius, ed., *L. Vitruvii Pollionis ad Cesarem Augustum*; "Io. Sulpitius Lectori Salutem," n.p., "non modo studiosis: sed reliquis hominibus." See also Manfredo Tafuri's introduction to "Cesare Cesariano e gli studi Vitruviani nel Quattrocento," in *Scritti rinascimentali di architettura,* ed. Analdo Bruschi, Corrado Maltese, Manfredo Tafuri, and Renato Bonelli (Milan: Il Polifilo, 1978), 387–458, on 394–398.

49 Claudio Sgarbi, ed., *Vitruvio ferrarese "De architectura": La prima versione illustrate* (Modena: Franco Cosimo Panini, 2004), esp. 11–20.

50 Vitruvius, *Ten Books on Architecture: The Corsini Incunabulum,* ed. Rowland, is a facsimile edition.

51 For Giovanni Giocondo, see esp. Lucia A. Ciapponi, "Appunti per una biografia di Giovanni Giocondo da Verona," *Italia Medioevale e umanistica* 4 (1961): 131–158; P. N. Pagliara, "Giovanni Giocondo da Verona (Fra Giocondo)," *Dizionario biografico degli Italiani,* 74 vols. (Rome: Istituto della Enciclopedia italiana, 1960–), 56:326–338; and Vincenzo Fontana, *Fra' Giovanni Giocondo Architetto 1433–c. 1515* (Vicenza: Neri Pozza Editore, 1988), 21–36 (for his work in Naples, including the Poggio Reale), 31–32 (for the drawings, which Fontana suggests were for two no longer extant redactions of his treatise (similar to Magliabechiana II, I, 141 in the Biblioteca Nazionale in Florence) for Alfonso, which Francesco left without having time to illustrate them, 51–80 (on Giocondo's work on fortification in the Veneto). For Francesco's exact whereabouts, see Flavia Cantatore, "Biografia cronologica di Francesco di Giorgio architetto," in *Francesco di Giorgio, architetto,* ed. Fiore and Tafuri, 412–413.

52 Fontana, *Fra' Giovanni Giocondo,* 15–20, citation on 16, "una conoscenza scientifica e precisa dell'antichità, per mezzo di vere e proprie campagne di misurazione, con schizzi presi dal vero e misurazioni quotate," and note 1, 17–18, for Giocondo's dedication to Lorenzo in which he mentions how he obtained the epigraphs in the collection. The location of Giocondo's three epigraphic collections are as follows: first redaction, dedicated to Lorenzo de Medici: Vatican City, Biblioteca Apostolica Vaticana, Vat. Lat. 10228; the second, dedicated to Ludovico Agnelli, archbishop of Cosenza: Florence, Biblioteca Nazionale Centrale, Cod. Magl. CL; and the third: Venice: Marciana: Marciano Latino cl. XIV, 171 = 4665.

53 Fontana, *Fra' Giovanni Giocondo,* 47–48. The collection of drawings and notes are in Florence, Biblioteca Laurenziana, Plut. 29.43. Another manuscript contains selections from mathematical texts and notes by Giocondo—Vatican City, Biblioteca Apostolica Vatiana, Vat. Lat. 4539. See esp. Lucia A. Ciapponi, "Disegni ed appunti di matematica in un codice di Fra Giocondo da Verona (Laur. 29, 43)," in *Vestigia: Studi in onore di Giuseppe Billanovich,* 2 vols., ed. Rino Avesani, Mirella Ferrari, Tino Foffano, Giuseppe Frasso, and Agostino Sottili (Rome: Edizioni di Storia e Letteratura, 1984), 1:181–196.

54 See Fontana, *Fra' Giovanni Giocondo*, 40–44 (for the Pont Notre Dame), 46–47 (lectures on Vitruvius), 51–80 (his work in the Veneto), and 74–76 (for the two Vitruvian editions).

55 Giovanni Giocondo, *M. Vitruvius per Iocundum solito castigatior factus cum figures et tabula . . .* (Venice: Ioannis de Tridino alias Tacuino, 1511); and Giovanni Giocondo, *M. Vitruvius et Frontinus a Jocundo revisi repurgatique quantum ex collatione licuit* (Florence: Filippo Giunta, 1513); "Iuliano Medicae frater Io. Iocundus. S.P.D.," n.p., "valeat artifex, et quantum liberalia studia mechanicis addant, quae perinde ut vivax spums sunt corpori." See esp. Lucia A. Ciapponi, "Fra Giocondo da Verona and His Edition of Vitruvius," *Journal of the Warburg and Courtauld Institutes* 47 (1984): 72–90, who emphasizes Giocondo's often correct emendations of the corrupt text and his efforts to make it understandable to practitioners; and Rowland, *Culture of the High Renaissance,* 176–178.

56 See Vincenzo Fontana and Paolo Morachiello, eds., *Vitruvio e Raffaello: Il "De architectura" di Vitruvio nella traduzione inedita di Fabio Calvo Ravennate* (Rome: Officina Edizioni, 1975). The copies are in Munich, Bayerische Staatsbibliothek, Cod. It. 37 (complete with Raphael's additions) and It. 37a. For Calvo, see Vincenzo Fontana, "Elementi per una biografia di M. Fabio Calvo Ravenneta," in *Vitruvio e Raffaello*, ed. Fontana and Morachiello, 45–61, which includes extensive discussion of the sources; and Philip J. Jacks, "Calvo, Marco Fabio," in *Dictionary of Art,* ed. Turner, 5:448; and R. Gualdo, "Fabio Calvo, Marco," *Dizionario biografico degli Italiani*, 43:723–727. And for Raphael's methods of measuring, Rowland, introduction to Vitruvius, *Ten Books on Architecture: The Corsini Incunabulum*, 17.

57 Vitruvius, De architectura libri dece [da Caesare Caesariano] ([Como]: G. da Ponte, 1521), fol. 91v, "lo grammatical opusculo di Donato"; "la sua natural furibondia." See esp. Maria Luisa Gatti Perer and Alessandro Rovetta, eds., Cesare Cesariano e il classicismo di primo Cinquecento (Milan: Vita e Pensiero 1996), for studies and Cesariano's translation and commentary. A succinct summary of Cesariano's life is Francesco Paolo Fiore, "Cesariano [Ciserano], Cesare," in Dictionary of Art, ed. Turner, 6:356–359, citation—"geometric discipline" on 357. See also Alessandro Rovetta, Elio Monducci, and Corrado Caselli, Cesare Cesariano e il Rinascimento a Reggio Emilia (Milan: Silvana, 2008).

58 Vitruvius, *De architectura libri dece [da Caesare Caesariano],*, fol. 91v, "per inspeculare e cognoscere varii ingenii e costumi de homini: conversando e studendo asai," "Qui male nati sunt impossibile est ut bene facere possint," "lo illuminatore de questa divina opera," "maxima utilitate et necessitate," and (for the allegory of his own life), fol. 92r. Virtually all of Cesariano's Aristotelian citations appear in a small sentence book in which excerpts from various classical authors including Aristotle are listed—*Propositiones Aristoteles* (Venice: [Georgius Arrivabenus], ca. 1490), fol. 27, making this his probable source.

59 Vitruvius, *De architectura libri dece [da Caesare Caesariano]*, fol. 46v, "maxime de li docti e sapienti litterati," "de li artifice quali asai hano insudato," "secundola loro bona professione utile e necessaria," "ben che alcune volte la fortuna qualcuno di quisti li afaticasse e li vexasse."

60 Vitruvius, *De architectura libri dece [da Caesare Caesariano]*, fol. 10r–12r, "non solum la Architectura: ma ciscune altre arte", "di opera seu fabricatione e di rationcinatione"; "ben calculate e considerate," "saper dire e fare," "quasi a magiore opportunita," "il parlare de quella operaria cosa con ratione" (10r), "per expositione trahendo il senso de la cosa como fano li periti magistri de qualche artificio che non solum con li dicti ma con li facti dimonstrano le arte per erudire li rudi operantii," "in questa vita nisi per causa de la tractatione" (11v), "che sano operare con le tractatione se perduceno a la elegantia . . . per essere allegati del suo sapere" (11v–12r), "non solum un pocho da li Praeceptori: ma da la natura" (12r).

61 Vitruvius, *De architectura libri dece [da Caesare Caesariano]*, fol. 2v, "convien fare ordine: acio si possa demonstrare le cose che sono al proposito de la formatione de qualunque cose intendemo operare," "La imaginatione de nostri pensieri sia per ordine dimonstrato per il sucessuro caso seu effecto: quale per rationcinatione con la experientia ne indica larte e la preformatione de le cose."

62 Vitruvius, *De architectura libri dece [da Caesare Caesariano]*, fol. 18r, "la excogitativa e effectrice e inventrice del operatione manual," "questa Machinatione intellectiva cum sia causa de la formatione de li instrumenti fabrili: seu artisti opportuni ad explicare lo effecto de qualunque cose che noi volemo perficere," "non solum in larte millitare bisogna questa ingeniosa scientia Mechanica ma in tute le liberale dimonstratione e operatione senza la qual niuno ornamento opportuno del mundo saria quasi possible per lo uso de la vita commune ne del vestire e altre infinite cose artiste necessarie che al uso humano e necessario."

63 Vitruvius, *De architectura libri dece [da Caesare Caesariano]*, fol. 162v, "praeclari Philosophi," "la contemplatione intesa," "le magne cogitatione," "uno ardente desiderio de produre in opera sensibile con le proprie mane quello che con la mente havevano rationcinato."

64 For the dispute between Cesariano and the editors, see Fiore, "Cesariano," 5, 359, and for an edition of the Madrid manuscript, Cesare Cesariano, *Volgarizzamento dei libri IX (capitol 7 e 8) e X di Vitruvio, "De architectura," secondo il manoscrito 9/2790 Secciòn de Cortes della Real Academia de la Historia, Madrid*, ed. Barbara Agosti (Pisa: Scuola Normale Superiore [1996]).

Chapter Four

1 Peter Galison, *Image and Logic: A Material Culture of Microphysics* (Chicago: University of Chicago Press, 1997), 781–844; Peter Galison, "Computer Simulations and the Trading Zone," in *The Disunity of Science: Boundaries, Contexts, and Power*, ed. Peter Galison and David J. Stump (Stanford, Calif.: Stanford University Press, 1996), 118–157; and Peter Galison, "Trading with the Enemy," in *Trading Zones and Interactional Expertise*, ed. Michael E. Gorman (Cambridge, Mass.: MIT Press, 2010), 25–52.

2 For a study of expertise in these centuries, see Eric H. Ash, *Power, Knowledge, and Expertise in Elizabethan England* (Baltimore: Johns Hopkins University Press, 2004).

3 For discussions of patronage in the history of science, see esp. Bruce T. Moran, ed., *Patronage and Institutions: Science, Technology, and Medicine at the European Court, 1500–1750* (Rochester, N.Y.: Boydell Press, 1991); and Mario Biagioli, *Galileo Courtier: The Practice of Science in the Culture of Absolutism* (Chicago: University of Chicago Press, 1993).

4 Marcel Mauss, *The Gift: Form and Reason for Exchange in Archaic Societies*, trans. W. D. Halls (New York: W. W. Norton, 1990). See also Pierre Bourdieu, *Outline of a Theory of Practice*, trans. Richard Nice (Cambridge: Cambridge University Press, 1977), esp. 4–5, which points out that if subjective gift giving were understood in the way Mauss describes it objectively, the practice would fall apart. See also Paula Findlen, "The Economy of Scientific Exchange in Early Modern Italy," in *Patronage and Institutions*, ed. Moran, 5–24.

5 To say that the individuals from diverse backgrounds came to lose some of their differences is not to suggest that they became the same. Although there are some instances where it is impossible to know which background a particular person came from, usually an individual maintained his own identity while adopting some of the values of the other group. For studies of some of the trading zones mentioned here, see Bruce T. Moran, "Courts and Academies," and Adrian Johns, "Coffeehouses and Print Shops," in *The Cambridge History of Science,* vol. 3: *Early Modern Science*, ed. Katharine Park and Lorraine Daston (Cambridge: Cambridge University Press, 2006), 251–271 and 320–340, respectively; and Rob Iliffe, "Material Doubts: Hooke, Artisan Culture, and the Exchange of Information in 1670s London," *British Journal for the History of Science* 28, no. 3 (1995): 285–318. For instrument makers' shops, see James A. Bennett, "Shopping for Instruments in Paris and London," in *Merchants and Marvels: Commerce, Science, and Art in Early Modern Europe*, ed. Pamela H. Smith and Paula Findlen (New York: Routledge, 2002), 370–395.

6 See Kelly DeVries, "Sites of Military Science and Technology," in *The Cambridge History of Science*, vol. 3: *Early Modern Science*, ed. Park and Daston, 306–319. For arsenals, see esp. Ennio Concina, ed., *Arsenali e città nell'occidente europeo* (Rome: La Nuova Italia Scientifica, 1987); and Ennio Concina, *L'Arsenale della Repubblica di Venezia* (Milan: Electa, 1984). For writings, see Maurice J. D. Cockle, *A Bibliography of Military Books up to 1642*, 2d ed. (London: Holland Press, 1957); Rainer Leng, *Ars belli: Deutsche taktische und kriegstechnische Bilderhandschriften und Traktate im 15. und 16. Jahrhundert*, 2 vols. (Wiesbaden: Reichert Verlag, 2002); and Martha D. Pollak, *Military Architecture, Cartography, and the Representation of the Early Modern European City: A Checklist of Treatises on Fortification in the Newberry Library* (Chicago: Newberry Library, 1991).

7 Kelly DeVries, *Medieval Military Technology* (Peterborough, Ont.: Broadview Press, 1992), 143–168; Bert S. Hall, *Weapons and Warfare in Renaissance Europe* (Baltimore: Johns Hopkins University Press, 1997), 105–200;

and Volker Schmidtchen, *Bombarden, Befestigungen, Büchsenmeister: Von den ersten Mauerbrechern des Spätmittelalters zur Belagerungsartillerie der Renaissance* (Düsseldorf: Droste, 1977), 1–42.

8 See Hall, *Weapons and Warfare*, 41–104; and Schmidtchen, *Bombarden*, 102–119.

9 See Erich Egg, *Das Handwerk der Uhr- und der Büchsenmacher in Tirol* (Innsbruck: Universitätsverlag Wagner, 1982), 183–201; and Erich Egg, *Der Tiroler Geschützguss, 1400–1600* (Innsbruck: Universitätsverlag Wagner, 1961), esp. 95–162.

10 Egg, *Der Tiroler Geschützguss;* Erich Egg, "From the Beginning to the Battle of Marignano – 1515," in *Guns: An Illustrated History of Artillery*, ed. Joseph Jobé (Greenwich, Conn.: New York Graphic Society, 1971), 9–36; Schmidtchen, *Bombarden*, 83–94.

11 Erich Egg, "From Marignano to the Thirty Years' War, 1515–1648," in *Guns: An Illustrated History of Artillery*, ed. Jobé, 37–54; and David Goodman, *Power and Penury: Government, Technology, and Science in Philip II's Spain* (Cambridge: Cambridge University Press, 1988), 88–150. For Hogge and the English context, see Edmund B. Teesdale, *Gunfounding in the Weald in the Sixteenth Century* (London: Trustees of the Royal Armouries, 1991); and Edmund B. Teesdale, *The Queen's Gunstonemaker: Being an Account of Ralph Hogge, Elizabethan Ironmaster and Gunfounder* (Seaford, Eng.: Lindel, 1984). For an introduction to the work of Georg Hartmann, see John P. Lamprey, "An Examination of Two Groups of Georg Hartmann Sixteenth-century Astrolabes and the Tables Used in Their Manufacture," *Annals of Science* 54, no. 2 (1997): 111–142.

12 Goodman, *Power and Penury*, 88–108; David Goodman, *Spanish Naval Power, 1589–1665: Reconstruction and Defeat* (Cambridge: Cambridge University Press, 1997); and Carla Rahn Phillips, *Six Galleons for the King of Spain: Imperial Defense in the Early Seventeenth Century* (Baltimore: Johns Hopkins University Press, 1986). For a classic account of the relationships of artillery and ships in the Mediterranean, see John Francis Guilmartin, Jr., *Gunpowder and Galleys: Changing Technology and Mediterranean Warfare at Sea in the Sixteenth Century* (Cambridge: Cambridge University Press, 1974). For an excellent account of Spanish efforts to acquire and monopolize new technical and scientific knowledge, see Marìa M. Portuondo, *Secret Science: Spanish Cosmography and the New World* (Chicago: University of Chicago Press, 2009).

13 A. Rupert Hall's view that there was no connection between scientific ballistics and the practices of gunners has been effectively challenged by Frances Willmoth, who shows that the ordnance office played an important role in sustaining traditions of practical mathematics and mechanics, which included studies of ballistics. See A. Rupert Hall, *Ballistics in the Seventeenth Century: A Study in the Relations of Science and War with Reference Principally to England* (Cambridge: Cambridge University Press, 1952); Richard W. Stewart, *The English Ordnance Office, 1585–1625: A Case Study in Bureaucracy* (Woodbridge, Suffolk: Boydel Press, 1996); and Frances Willmoth, *Sir Jonas Moore: Practical Mathematics and Restorations Science* (Woodbridge, Suffolk: Boydell Press,

1993); and for Tudor fortification, Steven A. Walton, "State Building through Building for the State: Foreign and Domestic Expertise in Tudor Fortification," in *Expertise: Practical Knowledge and the Early Modern State,* ed. Eric H. Ash, *Osiris,* 2d ser. 25 (2010): 66–84. See also Steven Johnston, "Making Mathematical Practice: Gentlemen, Practitioners, and Artisans in Elizabethan England" (Ph.D. diss., Cambridge University, 1994), http://www.mhs.ox.ac.uk/staff/saj/thesis/abstract.htm (accessed 17 May 2011), which contains much about artillery and fortification as well as other mathematical topics; and see Anthony Gerbino and Stephen Johnston, *Compass and Rule: Architecture as Mathematical Practice in England* (Oxford: Museum of the History of Science, 2009).

14 See esp. Giorgio Bellavitis, *L'Arsenale di Venezia: Storia di una grande struttura urbana* (Venice: Marsilio Editori, 1983); Concina, *L'Arsenale della Repubblica*; Robert C. Davis, *Shipbuilders of the Venetian Arsenal: Workers and Workplace in the Preindustrial City* (Baltimore: Johns Hopkins University Press, 1991); and Franco Rossi, "L'Arsenale: I quadri direttivi," in *Storia di Venezia: Dalle origini alla caduta della Serenissima,* vol. 5: *Il Rinascimento: Società ed economia,* ed. Alberto Tenenti and Ugo Tucci (Rome: Istituto della Enciclopedia Italiana, 1996), 593–639.

15 Mauro Bondioli, "Early Shipbuilding Records and the Book of Michael of Rhodes," in *The Book of Michael of Rhodes: A Fifteenth-Century Maritime Manuscript,* 3 vols., ed. Pamela O. Long, David McGee, and Alan M. Stahl (Cambridge, Mass.: MIT Press, 2009), 3:243–280, esp. 271–280; and Frederic Chapin Lane, *Venetian Ships and Shipbuilders of the Renaissance* (Baltimore: Johns Hopkins University Press, 1934), 56–59.

16 Pamela O. Long, David McGee, and Alan M. Stahl, eds., *The Book of Michael of Rhodes: A Fifteenth-Century Maritime Manuscript,* 3 vols. (Cambridge, Mass.: MIT Press, 2009), 1:211–217 and 2:272–281, for Michael's service record of his voyages. Based on his service record and other archival documents, Alan M. Stahl has written a detailed biography: "Michael of Rhodes: Mariner in Service to Venice," in *Book of Michael of Rhodes,* ed. Long, McGee, and Stahl, 3:35–98.

17 Raffaella Franci, "Mathematics in the Manuscript of Michael of Rhodes," in *Book of Michael of Rhodes,* ed. Long, McGee, and Stahl, 3:115–146; Franci shows that Michael was a good mathematician and worked the problems himself.

18 Piero Falchetta, "The Portolan of Michael of Rhodes," in *Book of Michael of Rhodes,* ed. Long, McGee, and Stahl, 3:193–210. Falchetta discusses the errors in the portolan and suggests that it was included not as a practical guide for navigators but to impress noble patrons and other nonskilled elite persons.

19 Dieter Blume, "The Use of Visual Images by Michael of Rhodes: Astrology, Christian Faith, and Practical Knowledge," in *Book of Michael of Rhodes,* ed. Long, McGee, and Stahl, 3:147–191.

20 Faith Wallis, "Michael of Rhodes and Time Reckoning: Calendar, Almanac, Prognostication," in *Book of Michael of Rhodes,* ed. Long, McGee, and Stahl, 3:281–319.

21 Blume, "Use of Visual Images," 3:177, suggests the context of reversal in a hierarchical society in which Michael created his image, a subversion of the heraldic code.

22 David McGee, "The Shipbuilding Text of Michael of Rhodes," and Bondioli, "Early Shipbuilding Records," in *Book of Michael of Rhodes*, ed. Long, McGee, and Stahl, 3:211–241 and 243–280, respectively.

23 Falchetta, "Portolan of Michael of Rhodes," 3:193–210, shows that the portolan contains numerous errors and that it could hardly have been meant as a guide for practical navigation. What he writes concerning portolan texts also could apply also to other parts of Michael's book: "They [portolans] were no longer necessary or at any rate useful instruments for going to sea. Rather they were texts distinguished by a certain degree of autonomy, whose contents only have the appearance of being technical. In the final analysis we can assert that these texts no longer maintained an instrumental function—or rather, their instrumental function seems to be much less significant than the new function they assumed, which belonged primarily within the symbolic sphere. They manifest both the transformation of the epistemological framework of knowledge related to navigation as well as the transformation of the more general system of cultural values. Thus they bring into focus, even within the maritime world, the ever-increasing value of the *libro* as testimony—as well as means—of knowledge" (210).

24 McGee, "Shipbuilding Text," 3: 237–241.

25 For more details, see Pamela O. Long, "Introduction: The World of Michael of Rhodes, Venetian Mariner," in *Book of Michael of Rhodes*, ed. Long, McGee, and Stahl, 3:1–33, esp. 15–20 and 30–31.

26 Nicholas of Cusa, *Opera omnia*, vol. 5: *Idiota: De sapientia, de mente, de staticis experimentis*, rev. ed., ed. Renata Steiger et al. from the edition of Ludwig Baur (Hamburg: Felix Meiner, 1983). There is an anonymous English translation, *The Idiot in Four Books* (London: William Leake, 1650); and a more recent English translation of *De Mente*—Nicholas of Cusa, *Idiota de Mente: The Layman: About Mind*, trans. Clyde Lee Miller (New York: Abaris, 1979).

27 See Ennio Concina, *Navis: L'umanismo sul mare (1470–1740)* (Turin: Giulio Einaudi, 1990); and Lane, *Venetian Ships*, 56–59, 64–71. For Vettor Vausto, see F. Piovan, "Fausto, Vittore," *Dizionario biografico degli Italiani* (Rome: Istituto della Enciclopedia Italiana, 1960–), 45:398–401; and N. G. Wilson, "Vettor Fausto, Professor of Greek and Naval Architect," in *The Uses of Greek and Latin*, ed. A. C. Dionisotti, Anthony Grafton, and Jill Kraye (London: Warburg Institute, 1988), 89–95.

28 See Concina, Navis; Lane, Venetian Ships, 64–71. For Galileo, the arsenal, and the military compass, see Galileo Galilei, Operations of the Geometric and Military Compass, 1606, trans. Stillman Drake (Washington, D.C.: Smithsonian Institution Press, 1978). And see Jürgen Renn and Matteo Valleriani, "Galileo and the Challenge of the Arsenal," Nuncius 16, no. 2 (2001): 481–503; and Matteo Valleriani, Galileo Engineer (Dordrecht: Springer, 2010), 27–38 (on the compass), 117–153 (on Galileo and the arsenal).

29 Vannoccio Biringuccio, *De la pirotechnia, 1540,* ed. Adriano Carugo (1540; Milan: Il Polifilo, 1977); and an English translation, Vannoccio Biringuccio, *The Pirotechnia of Vannoccio Biringuccio: The Classic Sixteenth-Century Treatise on Metals and Metallurgy,* trans. and ed. Cyril Stanley Smith and Martha Teach Gnudi, 2d ed. (1959; New York: Dover Books, 1990), 213–260 (bk. 6, 1–11).

30 Niccolò Tartaglia, *Nova scientia inventa da Nicolo Tartalea B.* (Venice: Stephano da Sabio, 1537); and Niccolò Tartaglia, *Quesiti et inventioni diverse de Nicolo Tartalea Brisciano* (Venice: Venturino Ruffinelli for N. Tartaglia, 1546). And see Gerhard Arend, *Die Mechanik des Niccolò Tartaglia im Kontext der zeitgenössischen Erkenntnis- und Wissenschaftstheorie* (Munich: Institut für Geschichte der Naturwissenschaften, 1998); Serafina Cuomo, "Shooting by the Book: Notes on Niccolò Tartaglia's 'Nova Scientia,'" *History of Science* 35 (June 1997): 155–188; and Mary J. Henninger-Voss, "How the 'New Science' of Cannons Shook Up the Aristotelian Cosmos," *Journal of the History of Ideas* 63 (July 2002): 371–397.

31 John U. Nef, "Mining and Metallurgy in Medieval Civilisation," in The Cambridge Economic History of Europe, vol. 2: Trade and Industry in the Middle Ages, ed. M. M. Postan and Edward Miller, assisted by Cynthia Postan, 2d ed. (Cambridge: Cambridge University Press, 1987), 691–761, 933–940; and see Ian Blanchard, Mining, Metallurgy and Minting in the Middle Ages, 3 vols. (Stuttgart: F. Steiner, 2001–2005), vols. 2 and 3. Mining scholarship includes numerous archivally based studies of local regions. See, for example, Angelika Westermann, Entwicklungsprobleme der Vorderösterreichischen Montanwirtschaft im 16. Jahrhundert: Eine verwaltungs', rechts-, wirtschafts-, und sozialgeschichtliche Studie als Vobereitung für einen multiperspektivischen Geschichtsunterricht (Idstein: Schulz-Kirchner, 1993); and Catherine Verna, Les mines et les forges des Cisterciens en Champagne méridionale et en Bourgogne du nord, XIIe–XVe siècle (Paris: Association pour l'Édition et la Diffusion del Études Historiques, 1995).

32 Nef, "Mining and Metallurgy," 723–746. For Agricola, see Georg Agricola, De re metallica libri XII (Basel: H. Frobenius and N. Episcopius, 1556); and the English translation with extensive technical notes, Agricola, De re metallica, trans. Herbert Clark Hoover and Lou Henry Hoover (1912; New York: Dover, 1950).

33 Nef, "Mining and Metallurgy," 723–746; and Blanchard, Mining and Metallurgy, 3:1071–1074. For coal production, see esp. Paul Benoit and Catherine Verna, eds., Le Charbon de terre en Europe occidentale avant l'usage industriel du coke, Proceedings of the XXth International Congress of History of Science (Liège, 20–26 July 1997) (Turnhout, Belgium: Brepols, 1999).

34 Philippe Braunstein, "Innovations in Mining and Metal Production in Europe in the Late Middle Ages," Journal of European Economic History" 12 (Winter 1983): 573–591; and Hermann Kellenbenz, The Rise of the European Economy: An Economic History of Continental Europe from the Fifteenth to the Eighteenth Century, ed. Gerhard Benecke (London: Weidenfeld and Nicolson, 1976), 85–88.

35 Braunstein, "Innovations in Mining," 587–591.

36 For these writings and further bibliography, see Pamela O. Long, "The Openness of Knowledge: An Ideal and Its Context in 16th-Century Writings on Mining and Metallurgy," *Technology and Culture* 32 (April 1991): 318–355.

37 See ibid., and Pamela O. Long, *Openness, Secrecy, Authorship: Technical Arts and the Culture of Knowledge from Antiquity to the Renaissance* (Baltimore: Johns Hopkins University Press, 2001), 176–191. For Biringuccio, see note 29.

38 For hydraulic projects, see David Karmon, "Restoring the Ancient Water Supply System in Renaissance Rome: The Popes, the Civic Administration, and the Acqua Vergine," *Aqua urbis Romae: The Waters of the City of Rome*, http://www.iath.virginia.edu/waters (accessed 19 May 2011); Pamela O. Long, "Hydraulic Engineering and the Study of Antiquity: Rome, 1557–70," *Renaissance Quarterly* 61 (Winter 2008): 1098–1138; and Katherine Wentworth Rinne, *The Waters of Rome: Aqueducts, Fountains, and the Birth of the Baroque City* (New Haven, Conn.: Yale University Press, 2010). For the obelisks, see Brian A. Curran, Anthony Grafton, Pamela O. Long, and Benjamin Weiss, *Obelisk: A History* (Cambridge, Mass.: Burndy Library Publications, MIT Press, 2009).

39 See especially the essays in Antonella Romano, ed., *Rome et la science moderne entre Renaissance et Lumières* (Rome: École Française de Rome, 2008); and Maria Pia Donato and Jill Kraye, eds., *Conflicting Duties: Science, Medicine, and Religion in Rome, 1550–1750* (London: Warburg Institute, 2009).

40 Long, "Hydraulic Engineering," 1101–1103 .

41 Ibid., 1103–1109. For Bacci's flood writings, see Andrea Bacci, *Del Tevere: Della Natura et bonta dell'Acque e Delle Inondationi Libri II* (Rome: Vincenzo Luchino, 1558); Bacci, *Del Tevere di m. Andrea Bacci medico et filosofo, libri tre* (Venice: [n.p.], 1576); and Bacci, *Del Tevere libro quarto* (Rome: Stampatori Camerali, 1599). The handwritten tract is in the flyleaves of the 1976 edition of *Del Tevere,* in the Vatican Library in Vatican City: BAV, shelf no. Aldine II 98.

42 Antonio Trevisi, *Fondamento del edifitio nel quale si tratta con la santita de N.S. Pio Papa IIII Sopra la innondatione del Fiume* (Rome: Antonio Blado, 1560). For raising the ancient ship, see esp. Concina, *Navis,* 4–21. For Alberti's attempt to raise the ship, see esp. Anthony Grafton, *Leon Battista Alberti: Master Builder of the Italian Renaissance* (New York: Hill and Wang, 2000), 248–253.

43 See Long, "Hydraulic Engineering," 1113–1116; and Jessica Maier, "Mapping Past and Present: Leonardo Bufalini's Plan of Rome (1551)," *Imago Mundi* 59, no. 1 (2007): 1–23.

44 Long, "Hydraulic Engineering," 1119–1123; and Luca Peto, *Discorso di Luca Peto intorno alla cagione della Eccessiva Inondatione del Tevere in Roma, et modo in parte di soccorrervi* (Rome: Giuseppe degl'Angeli, 1573); and Peto, *De mensuris et ponderibus Romanis et Graecis* (Venice: [P. Manutius], 1573).

45 Curran, Grafton, Long, and Weiss, *Obelisk.* See also Brian Curran, *The Egyptian Renaissance: The Afterlife of Ancient Egypt in Early Modern Italy* (Chicago: University of Chicago Press, 2007); and see Giovanni Cipriani, *Gli*

obelischi egizi: Politica e cultura nella Roma barocca (Florence: Leo S. Olschki, 1993).

46 Curran, Grafton, Long, and Weiss, *Obelisk*, 107–109; and Camillo Agrippa, *Trattato di Camillo Agrippa Milanese De trasportar la guglia in su la piazza di San Pietro* (Rome: Francesco Zanetti, 1583).

47 Francesco Masini, *Discorso di Francesco Masini sopra un modo nuovo, facile, e reale, di trasportar su la Piazza di San Pietro la guglia, ch'è in Roma, detta di Cesare* (Cesena: Bartolomeo Raverij, 1586).

48 See Curran, Grafton, Long, and Weiss, *Obelisk,* 102–139, for an account of the competition and successful move. For Fontana's treatise, see Domenico Fontana, *Della trasportatione dell'obelisco vaticano et delle fabriche di nostro signore Papa Sisto V, fatte dal cavallier Domenico Fontana, architetto di Sua Santita* (Rome: Domenico Basa, 1590). There is a modern edition with a useful introduction and notes—Domenico Fontana, *Della trasportatione dell'obelisco vaticano, 1590,* ed. Adriano Carugo, with an introduction by Paolo Portoghese (Milan: Il Polifilo, 1978); and an English translation— Domenico Fontana, *Della trasportatione dell'obelisco vaticano,* ed. Ingrid D. Rowland, trans. David Sullivan (Oakland, Calif.: Octavo, 2002).

49 Filippo Pigafetta, *Discorso di M. Filippo Pigafetta d'intorno all'historia della Aguglia, et alla ragione del muoverla* (Rome: Bartolomeo Graffi, 1586).

50 Michele Mercati, *De gli obelischi di Roma* (Rome: Domenico Basa, 1589); and a modern edition— Michele Mercati, *Gli obelischi di Roma,* ed. Gianfranco Cantelli (Bologna: Cappelli Editore, 1981).

51 Curran, Grafton, Long, and Weiss, *Obelisk,* esp. 116–134. For the interest of humanists, antiquarians, and others in the Egyptian past as it was manifest in Rome, see Cipriani, *Gli obelischi egizi,* and Curran, *Egyptian Renaissance.*

52 Andreas Beyer, "Palladio, Andrea [Gondola, Andrea di Pietro della]," in *The Dictionary of Art,* ed. Turner, 23:861–872; and see esp. Guido Beltramini and Howard Burns, eds., *Palladio* (Venice: Marsilio, 2008); Bruce Boucher, *Andrea Palladio: The Architect in His Time* (New York: Abbeville Press, 1994). Also useful is a timeline in Center for Palladian Studies in America, "Palladio's Life and World: A Timeline," http://www.palladiancenter.org/timeline-Palladio.html (accessed 19 May 2011). And see Alina A. Payne, *The Architectural Treatise in the Italian Renaissance: Architectural Invention, Ornament, and Literary Culture* (Cambridge: Cambridge University Press, 1999), 170–213.

53 In addition to the references in note 52, see Vaughan Hart and Peter Hicks, *Palladio's Rome: A Translation of Andrea Palladio's Two Guidebooks to Rome* (New Haven, Conn.: Yale University Press, 2006), xiv–liii. For Daniele Barbaro, see esp. G. Alberigo, "Barbaro, Daniele Matteo Alvise," in *Dizionario biografico degli Italiani,* 6:89–95; and Manfredo Tafuri, "Daniele Barbaro e la cultura scientifica veneziana del'500," in *Cultura, scienze e technica nella Venezia del cinquecento: Giovan Battista Benedetti e il suo tempo* (Venice: Istituto Veneto di Scienze, Lettere ed Arti, 1987), 55–81. His treatise on perspective is Daniele Barbaro, *La pratica della perspettiva di Monsignor Daniel Barbaro* (Venice: Camillo and Rutilio Borgominieri Fratelli, 1569).

54 For the Villa Maser, see Douglas Lewis, "Maser, Villa Barbaro," in *Dictionary of Art*, ed. Turner, 20:545–547. For Barbaro's Vitruvian commentary, see Daniele Barbaro, *I dieci libri dell'architettura tradotti et commentati da monsignor Barbaro* (Venice: F. Marcolini, 1556); and Daniele Barbaro, *I dieci libri dell'architettura tradotti e commentati da Daniele Barbaro, 1567*, ed. Manfredo Tafuri and Manuela Morresi (Milan: Il Polifilo, 1987). For Palladio's treatise on architecture, see Andrea Palladio, *I quattro libri dell'architettura di Andrea Palladio* (Venice: Dominico de' Franceschi, 1570); and Palladio, *The Four Books on Architecture*, trans. Robert Tavernor and Richard Schofield (Cambridge, Mass.: MIT Press, 1997). See also Hart and Hicks, *Palladio's Rome*.

55 For his topographical work, see Cipriano Piccolpasso, *Le piante et i ritratti delle città e terre dell' Umbria sottoposte al governo di Perugia*, ed. Giovanni Cecchini (Rome: Istituto Nazionale d'Archeologia e Storia dell'Arte, 1963). And see Piccolpasso, *The Three Books of the Potter's Art: A Facsimile of the Manuscript in the Victoria and Albert Museum, London*, 2 vols., trans. and introduced by Ronald Lightbown and Alan Caiger-Smith (London: Scholar Press, 1980), 2:105.

56 See Gerd Spies, ed., *Technik der Steingewinnung und der Flussschiffahrt in Harzvoland in früher Neuzeit* (Braunschweig: Waisenhaus, 1992); and for Julius's mining activities, see Hans-Joachim Kraschewski, *Wirtschaftspolitik im deutschen Territorialstaat des 16. Jahrhunderts: Herzog Julius von Braunschweig-Wolfenbüttel (1528–1589)* (Cologne: Böhlau Verlag, 1978), 151–165.

57 See Galison, "Trading with the Enemy," esp. 41–44.

Conclusion

1 For the Aristotelian framework, see Pamela O. Long, *Openness, Secrecy, Authorship: Technical Arts and the Culture of Knowledge from Antiquity to the Renaissance* (Baltimore: Johns Hopkins University Press, 2001), esp. 2–3 and 23–24. The status of both labor and the mechanical arts rose during the medieval period. See especially George Ovitt Jr., *The Restoration of Perfection: Labor and Technology in Medieval Culture* (New Brunswick, N.J.: Rutgers University Press, 1987), and Elspeth Whitney, *Paradise Restored: The Mechanical Arts from Antiquity through the Thirteenth Century. Transactions of the American Philosophical Society*, n.s., 80, pt. 1 (1990).

2 Lissa Roberts and Simon Schaffer, preface to, and Lissa Roberts, introduction to *The Mindful Hand: Inquiry and Invention from the Late Renaissance to Early Industrialisation*, ed. Lissa Roberts, Simon Schaffer, and Peter Dear (Amsterdam: Koninklijke Nederlandse Akademie van Wetenschappen, 2007), xiii–xxvii, esp. xiv–xv, and 1–8.

3 Mary Henninger-Voss, "Comets and Cannonballs: Reading Technology in a Sixteenth-Century Library"; Pamela H. Smith, "In a Sixteenth-Century Goldsmith's Workshop"; and Fokko Jan Dijksterhuis, "Constructive Thinking: A Case for Dioptrics"—all in *The Mindful Hand*, ed. Roberts, Schaffer, and Dear, 10–31, 32–57, and 58–82, respectively.

4 This conclusion dissents from Zilsel's view that explicitly separates humanists from "superior artisans"—Edgar Zilsel, "The Methods of Humanism," in Zilsel, *The Social Origins of Modern Science*, ed. Diederick Raven, Wolfgang Krohn, and Robert S. Cohen (Dordrecht: Kluwer Academic, 2000), 22–64.

5 For example, see Pamela H. Smith, *The Body of the Artisan: Art and Experience in the Scientific Revolution* (Chicago: University of Chicago Press, 2004), esp. 182–236: "Artisanal bodily experience was absorbed into the work of the natural philosopher at the same time that the artisan himself was excised from it" (186). For a nuanced discussion of the issue of the changing status and activities of practitioners in the light of their influence on experimental philosophy, see Bert de Munck, "Corpses, Live Models, and Nature: Assessing Skills and Knowledge before the Industrial Revolution (Case: Antwerp)," *Technology and Culture* 51 (April 2010): 332–356.

Bibliography

Agricola, Georg. *De re metallica*. Translated by Herbert Clark Hoover and Lou Henry Hoover. 1912. New York: Dover, 1950.

————. *De re metallica libri XII*. Basel: H. Frobenius and N. Episcopius, 1556.

Agrimi, Jole, and Chiara Crisciani. "Per una ricerca su *experimentum-experimenta*: Reflessione epistemologica e tradizione medica (secoli XIII–XV)." In *Presenza del lessico greco e latino nelle lingue contemporanee*, edited by Pietro Janni and Innocenzo Mazzini, 9–49. Macerata: Università degli Studi di Macerata, 1990.

Agrippa, Camillo. *Trattato di Camillo Agrippa Milanese De trasportar la guglia in su la piazza di San Pietro*. Rome: Francesco Zanetti, 1583.

Alberigo, G. "Barbaro, Daniele Matteo Alvise." *Dizionario biografico degli Italiani*, 6:89–95.

Alberti, Leon Battista. *De re aedificatoria*. Florence: Nicolaus Laurentii, [1486].

————. *L'architettura [De re aedificatoria]*. 2 vols. Edited and translated by Giovanni Orlandi. Introduction and notes by Paolo Portoghesi. Milan: Il Polifilo, 1966.

————. *On the Art of Building in Ten Books*. Translated by Joseph Rykwert, Neil Leach, and Robert Tavernor. Cambridge, Mass.: MIT Press, 1988.

————. *On Painting*. Introduction and notes by Martin Kemp. Translated by Cecil Grayson. London: Penguin Books, 1991.

Arend, Gerhard. *Die Mechanik des Niccolò Tartaglia im Kontext der zeitgenössischen Erkenntnis- und Wissenschaftstheorie*. Munich: Institut für Geschichte der Naturwissenschaften, 1998.

Aristotle. *Physics*. Translated by R. P. Hardie and R. K. Gaye. In *The Complete Works of Aristotle*, rev. ed., 2 vols., edited by Jonathan Barnes, 1:315–446. Princeton, N.J.: Princeton University Press, 1984.

Ash, Eric H. *Power, Knowledge, and Expertise in Elizabethan England*. Baltimore: Johns Hopkins University Press, 2004.

Bacci, Andrea. *Del Tevere: Della Natura et bonta dell'Acque e Delle Inondationi Libri II*. Rome: Vincenzo Luchino, 1558.

————. *Del Tevere di m. Andrea Bacci medico et filosofo, libri tre*. Venice: [n.p.], 1576.

————. *Del Tevere libro quarto*. Rome: Stampatori Camerali, 1599.

Barbaro, Daniele. *I dieci libri dell'architettura tradotti e commentati da Daniele Barbaro, 1567*. Edited by Manfredo Tafuri and Manuela Morresi. Milan: Il Polifilo, 1987.

————. *I dieci libri dell'architettura tradotti e commentati da monsignor Barbaro*. Venice: F. Marcolini, 1556.

————. *La pratica della perspettiva di Monsignor Daniel Barbaro*. Venice: Camillo and Rutilio Borgominieri Fratelli, 1569.

Barolini, Helen. *Aldus and His Dream Book*. New York: Italica Press, 1992.

Barrera-Osorio, Antonio. *Experiencing Nature: The Spanish American Empire and the Early Scientific Revolution.* Austin: University of Texas Press, 2006.

Bellavitis, Giorgio. *L'Arsenale di Venezia: Storia di una grande struttura urbana.* Venice: Marsilio Editori, 1983.

Beltramini, Guido, and Howard Burns, eds. *Palladio.* Venice: Marsilio, 2008.

Beltramini, Maria. "Francesco Filelfo e il Filarete: Nuovi contributi alla storia dell'amicizia fra il letterato e l'architetto nella Milano sforzesca." *Annali della Scuola Normale Superiore di Pisa, Classe di Lettere e Filosofia* ser. 4, no. 1 (1996): 119–125.

Bennett, James A. "The Mechanical Arts." In *The Cambridge History of Science,* vol. 3: *Early Modern Science,* edited by Park and Daston, 673–695.

———. "The Mechanics' Philosophy and the Mechanical Philosophy." *History of Science* 24 (March 1986): 1–28.

———. "Shopping for Instruments in Paris and London." In *Merchants and Marvels,* edited by Smith and Findlen, 370–395.

Benoit, Paul, and Catherine Verna, eds. *Le Charbon de terre en Europe occidentale avant l'usage industriel du coke.* Proceedings of the XXth International Congress of History of Science (Liège, 20–26 July 1997). Turnhout, Belgium: Brepols, 1999.

Bensaude-Vincent, Bernadette, and William R. Newman, eds. *The Artificial and the Natural: An Evolving Polarity.* Cambridge, Mass.: MIT Press, 2007.

Betts, Richard J. "On the Chronology of Francesco di Giorgio's Treatises: New Evidence from an Unpublished Manuscript." *Journal of the Society of Architectural Historians* 36 (March 1977): 3–14.

Beyer, Andreas. "Palladio, Andrea [Gondola, Andrea di Pietro della]." *Dictionary of Art,* edited by Turner, 23:861–872.

Biagioli, Mario. *Galileo, Courtier: The Practice of Science in the Culture of Absolutism.* Chicago: University of Chicago Press, 1993.

Biffi, Marco. "Introduzione." In Francesco di Giorgio Martini, *La traduzione del "De architectura,"* XI–CXVII.

Biringuccio, Vannoccio. *De la pirotechnia, 1540.* Edited by Adriano Carugo. 1540. Milan: Il Polifilo, 1977.

———. *The Pirotechnia of Vannoccio Biringuccio: The Classic Sixteenth-Century Treatise on Metals and Metallurgy.* Translated and edited by Cyril Stanley Smith and Martha Teach Gnudi. 2d ed. 1959. New York: Dover Books, 1990.

Blair, Ann. "Natural Philosophy." In *The Cambridge History of Science,* vol. 3: *Early Modern Science,* edited by Park and Daston, 365–406.

Blanchard, Ian. *Mining, Metallurgy, and Minting in the Middle Ages.* 3 vols. Stuttgart: F. Steiner, 2001–2005.

Blume, Dieter. "The Use of Visual Images by Michael of Rhodes: Astrology, Christian Faith, and Practical Knowledge." In *Book of Michael of Rhodes,* edited by Long, McGee, and Stahl, 3:147–191.

Bondioli, Mauro. "Early Shipbuilding Records and the Book of Michael of Rhodes." In *Book of Michael of Rhodes,* edited by Long, McGee, and Stahl, 3:243–280.

Borkenau, Franz. *Der Übergang vom feudalen zum bürgerlichen Weltbild: Studien zur Geschichte der Philosophie der Manufakturperiode.* 1934. New York: Arno Press, 1975.

———. "The Sociology of the Mechanistic World-Picture." Translated by Richard W. Hadden. *Science in Context* 1 (March 1987): 109–127.

———. "Zur Soziologie des mechanistischen Weltbildes." *Zeitschrift für Sozialforschung* 1.3 (1932): 311–355.

Borsi, Stefano. *Leon Battista Alberti e l'antichità Romana.* Florence: Edizioni Polistampa, 2004.

———. *Leon Battista Alberti e Roma.* Florence: Edizioni Polistampa, 2003.

Bottomore, Tom. "Austro-Marxism." In *Dictionary of Marxist Thought,* edited by Bottomore, 36–39.

———, ed. *A Dictionary of Marxist Thought.* Cambridge, Mass.: Harvard University Press, 1983.

Bottomore, Tom, and Patrick Goode, eds. and trans. *Austro-Marxism.* Oxford: Clarendon Press, 1978.

Boucher, Bruce. *Andrea Palladio: The Architect in His Time.* New York: Abbeville Press, 1994.

Bourdieu, Pierre. *Outline of a Theory of Practice.* Translated by Richard Nice. Cambridge: Cambridge University Press, 1977.

Bowker, Geof, and Bruno Latour. "A Booming Discipline Short of Discipline: (Social) Studies of Science in France." *Social Studies of Science* 17 (November 1987): 715–748.

Braudel, Fernand. *The Structures of Everyday Life: Civilization and Capitalism, 15th–18th Century.* Translated by Siân Reynolds. 3 vols. New York: Harper and Row, 1981.

Braunstein, Philippe. "Innovations in Mining and Metal Production in Europe in the Late Middle Ages." *Journal of European Economic History* 12 (Winter 1983): 573–591.

Brown, Patricia Fortini. *Private Lives in Renaissance Venice: Art, Architecture, and the Family.* New Haven, Conn.: Yale University Press, 2004.

Butterwegge, Christoph. *Austromarxismus und Staat: Politiktheorie und Praxis der österreichischen Sozialdemokratie zwischen den beiden Weltkriegen.* Marburg: Verlag Arbeit und Gesellschaft GmbH, 1991.

Callebat, Louis, ed. and trans. *Vitruve De L'architecture, Livre VIII.* Paris: Société D'Édition "Les Belle Lettres," 1973.

Callebat, Louis, and Philippe Fleury, eds. and trans. *Vitruve De L'architecture, Livre X.* Paris: Société D'Édition "Les Belle Lettres," 1986.

Cantatore, Flavia. "Biografia cronologica di Francesco di Giorgio architetto." In *Francesco di Giorgio, architetto,* edited by Fiore and Tafuri, 412–413.

Carlino, Andrea. *Books of the Body: Anatomical Ritual and Renaissance Learning.* Translated by John Tedeschi and Anne C. Tedeschi. Chicago: University of Chicago Press, 1999.

Carpo, Mario, and Francesco Furlan, eds. *Leon Battista Alberti's Delineation of the City of Rome (Descriptio urbis Romae).* Critical edition by Jean-Yves

Boriaud and Francesco Furlan, trans. Peter Hicks. Tempe: Arizona Center for Medieval and Renaissance Studies, 2007.

Cartwright, Nancy, Jordi Cat, Lola Fleck, and Thomas E. Uebel. *Otto Neurath: Philosophy between Science and Politics.* Cambridge: Cambridge University Press, 1996.

Casella, Maria T., and Giovanni Pozzi. *Francesco Colonna: Biografia e opera.* 2 vols. Padua: Editrice Antenore, 1959.

Cat, Jordi, Nancy Cartwright, and Hasok Chang. "Otto Neurath: Politics and the Unity of Science." In *The Disunity of Science,* edited by Galison and Stump, 347–369.

Celenza, Christopher S. *The Lost Italian Renaissance: Humanists, Historians, and Latin's Legacy.* Baltimore: Johns Hopkins University Press, 2004.

Center for Palladian Studies in America. "Palladio's Life and World: A Timeline." http://www.palladiancenter.org/timeline-Palladio.html.

Cesariano, Cesare. *Volgarizzamento dei libri IX (capitol 7 e 8) e X di Vitruvio, "De architectura," secondo il manoscrito 9/2790 Secciòn de Cortes della Real Academia de la Historia, Madrid.* Edited by Barbara Agosti. Pisa: Scuola Normale Superiore [1996].

Chartier, Roger. *The Order of Books: Readers, Authors, and Libraries in Europe between the Fourteenth and Eighteenth Centuries.* Translated by Lydia G. Cochrane. Stanford, Calif.: Stanford University Press, 1994.

Chilvers, C. A. J. "The Dilemmas of Seditious Men: The Crowther-Hessen Correspondence in the 1930s." *British Journal for the History of Science* 36 (December 2003): 417–435.

Ciapponi, Lucia A. "Appunti per una biografia di Giovanni Giocondo da Verona." *Italia Medioevale e umanistica* 4 (1961): 131–158.

———. "Disegni ed appunti di matematica in un codice di Fra Giocondo da Verona (Laur. 29, 43)." In *Vestigia: Studi in onore di Giuseppe Billanovich,* 2 vols., edited by Rino Avesani et al., 1:181–196. Rome: Edizioni di Storia e Letteratura, 1984.

———. "Fra Giocondo da Verona and His Edition of Vitruvius." *Journal of the Warburg and Courtauld Institutes* 47 (1984): 72–90.

Cipriani, Giovanni. *Gli obelischi egizi: Politica e cultura nella Roma Barocca.* Florence: Leo S. Olschki, 1993.

Clark, Stuart. *Vanities of the Eye: Vision in Early Modern European Culture.* Oxford: Oxford University Press, 2007.

Cockle, Maurice J. D. *A Bibliography of Military Books up to 1642.* 2d ed. London: Holland Press, 1957.

Cohen, H. Floris. *The Scientific Revolution: A Historiographical Inquiry.* Chicago: University of Chicago Press, 1994.

Cohen, I. Bernard. "The Publication of *Science, Technology and Society*: Circumstances and Consequences." *Isis* 79 (December 1988): 571–582.

———. *Revolution in Science.* Cambridge, Mass.: Belknap Press of Harvard University Press, 1985.

Cole, Thomas. *Democritus and the Sources of Greek Anthropology.* Chapel Hill, N.C.: American Philological Association, 1967.

Colonna, Francesco. *Hypnerotomachia Poliphili*. 2 vols. Edited by Giovanni Pozzi and Lucia A. Ciapponi. Padua: Editrice Antenore, 1964.

———. *Hypnerotomachia Poliphili: The Strife of Love in a Dream*. Translated by Joscelyn Godwin. New York: Thames and Hudson, 1999.

Concina, Ennio. *L'Arsenale della Repubblica di Venezia*. Milan: Electa, 1984.

———. *Navis: L'umanismo sul mare (1470–1740)*. Turin: Giulio Einaudi, 1990.

———, ed. *Arsenali e città nell'occidente europeo*. Rome: La Nuova Italia Scientifica, 1987.

Cook, Harold J. *Matters of Exchange: Commerce, Medicine, and Science in the Dutch Golden Age*. New Haven, Conn.: Yale University Press, 2007.

Cooper, Alix. *Inventing the Indigenous: Local Knowledge and Natural History in Early Modern Europe*. Cambridge: Cambridge University Press, 2007.

Cunningham, Andrew. *The Anatomical Renaissance: The Resurrection of Anatomical Projects of the Ancients*. Aldershot, Eng.: Scolar Press, 1997.

Cuomo, Serafina. "Shooting by the Book: Notes on Niccolò Tartaglia's 'Nova Scientia.'" *History of Science* 35 (June 1997): 155–188.

Curran, Brian A. *The Egyptian Renaissance: The Afterlife of Ancient Egypt in Early Modern Italy*. Chicago: University of Chicago Press, 2007.

Curran, Brian A., Anthony Grafton, Pamela O. Long, and Benjamin Weiss. *Obelisk: A History*. Cambridge, Mass.: Burndy Library Publications, MIT Press, 2009.

Dahms, Hans-Joachim. "Edgar Zilsels Projekt 'The Social Roots of Science' und seine Beziehungen zur Frankfurter Schule." In *Wien-Berlin-Prag*, edited by Haller and Stadler, 474–500.

D'Amico, John F. *Renaissance Humanism in Papal Rome: Humanists and Churchmen on the Eve of the Reformation*. Baltimore: Johns Hopkins University Press, 1983.

Daston, Lorraine. "The Nature of Nature in Early Modern Europe." *Configurations* 6 (Spring 1998): 149–172.

Daston, Lorraine, and Katharine Park. *Wonders and the Order of Nature, 1150–1750*. New York: Zone Books, 1998.

Davies, Paul, and David Hemsoll. "Alberti, Leon Battista." In *Dictionary of Art*, edited by Turner, 1:555–569.

Davis, Robert C. *Shipbuilders of the Venetian Arsenal: Workers and Workplace in the Preindustrial City*. Baltimore: Johns Hopkins University Press, 1991.

Dear, Peter. *Discipline and Experience: The Mathematical Way in the Scientific Revolution*. Chicago: University of Chicago Press, 1995.

———. "The Meanings of Experience." In *The Cambridge History of Science*, vol. 3: *Early Modern Science*, edited by Park and Daston, 106–131.

———. *Revolutionizing the Sciences: European Knowledge and Its Ambitions, 1500–1700*. Princeton, N.J.: Princeton University Press, 2001.

———. "What Is the History of Science the History Of?: Early Modern Roots of the Ideology of Modern Science." *Isis* 96 (September 2005): 390–406.

Dennis, Michael Aaron. "Historiography of Science: An American Perspective." In *Science in the Twentieth Century*, edited by John Krige and Dominique Pestre, 1–26. Amsterdam: Harwood Academic, 1997.

DeVries, Kelly. *Medieval Military Technology*. Peterborough, Ont.: Broadview Press, 1992.

———. "Sites of Military Science and Technology." In *The Cambridge History of Science*, vol. 3: *Early Modern Science*, edited by Park and Daston, 306–319.

DeVun, Leah. *Prophecy, Alchemy, and the End of Time: John of Rupescissa in the Late Middle Ages*. New York: Columbia University Press, 2009.

Dijksterhuis, Fokko Jan. "Constructive Thinking: A Case for Dioptrics." In *The Mindful Hand*, edited by Roberts, Schaffer, and Dear, 58–82.

Dinsmoor, William Bell. "The Literary Remains of Sebastiano Serlio." *Art Bulletin* 24 (March 1942): 55–91.

Dizionario biografico degli Italiani. 75+ vols. Rome: Istituto della Enciclopedia Italiana, 1960–.

Dobbs, Betty Jo Teeter. *The Janus Faces of Genius: The Role of Alchemy in Newton's Thought*. Cambridge: Cambridge University Press, 1991.

Donahue, William. "Astronomy." In *The Cambridge History of Science*, vol. 3: *Early Modern Science*, edited by Park and Daston, 562–595.

Donato, Maria Pia, and Jill Kraye, eds. *Conflicting Duties: Science, Medicine, and Religion in Rome, 1550–1750*. London: Warburg Institute, 2009.

Dvořak, Johann. *Edgar Zilsel und die Einheit der Erkenntnis*. Vienna: Löcker Verlag, 1981.

———. "Otto Neurath and Adult Education: Unity of Science, Materialism and Comprehensive Enlightenment." In *Rediscovering the Forgotten Vienna Circle*, edited by Uebel, 265–274.

Dwyer, Eugene, Peter Kidson, and Pier Nicola Pagliara. "Vitruvius." In *Dictionary of Art*, edited by Turner, 32:632–643.

Eamon, William. *Science and the Secrets of Nature: Books of Secrets in Medieval and Early Modern Culture*. Princeton, N.J.: Princeton University Press, 1994.

Edgerton, Samuel Y., Jr. *The Renaissance Rediscovery of Linear Perspective*. New York: Basic Books, 1975.

Egg, Erich. *Das Handwerk der Uhr- und der Büchsenmacher in Tirol*. Innsbruck: Universitätsverlag Wagner, 1982.

———. *Der Tiroler Geschützguss, 1400–1600*. Innsbruck: Universitätsverlag Wagner, 1961.

———. "From the Beginning to the Battle of Marignano—1515." In *Guns*, edited by Joseph Jobé, 9–36.

———. "From Marignano to the Thirty Years' War, 1515–1648." In *Guns*, edited by Joseph Jobé, 37–54.

Eisenstein, Elizabeth. *The Printing Press as an Agent of Change: Communications and Cultural Transformations in Early Modern Europe*. 2 vols. Cambridge: Cambridge University Press, 1979.

Elias, Norbert. *The Court Society*. Translated by Edmund Jephcott. New York: Pantheon Books, 1983.

Enebakk, Vidar. "Lilley Revisited: Or Science and Society in the Twentieth Century." *British Journal for the History of Science* 42 (December 2009): 563–593.

Epstein, Steven A. *An Economic and Social History of Later Medieval Europe, 1000–1500.* Cambridge: Cambridge University Press, 2009.

———. *Wage Labor and Guilds in Medieval Europe.* Chapel Hill: University of North Carolina Press, 1991.

Epstein, S. R. *Freedom and Growth: The Rise of States and Markets in Europe, 1300–1750.* New York: Routledge, 2000.

Falchetta, Piero. "The Portolan of Michael of Rhodes." In *Book of Michael of Rhodes,* edited by Long, McGee, and Stahl, 3:193–210.

Filarete (Antonio Averlino). *Filarete's Treatise on Architecture.* 2 vols. Edited and translated by John R. Spencer. New Haven, Conn.: Yale University Press, 1965.

———. *Trattato di Architectura.* 2 vols. Edited by Anna Maria Finoli and Liliana Grassi. Milan: Edizioni il Polifilo, 1972.

Findlen, Paula. "The Economy of Scientific Exchange in Early Modern Italy." In *Patronage and Institutions,* edited by Moran, 5–24.

———. *Possessing Nature: Museums, Collecting, and Scientific Culture in Early Modern Italy.* Berkeley: University of California Press, 1994.

Fiore, Francesco Paolo. "Cesariano [Ciserano], Cesare." In *Dictionary of Art,* edited by Turner, 6:356–359.

Fiore, Francesco Paolo, and Manfredo Tafuri, eds. *Francesco di Giorgio architetto.* Milan: Electa, 1993.

Fontana, Domenico. *Della trasportatione dell'obelisco vaticano.* Edited by Ingrid D. Rowland. Translated by David Sullivan. Oakland, Calif.: Octavo, 2002.

———. *Della trasportatione dell'obelisco vaticano, 1590.* Edited by Adriano Carugo with an introduction by Paolo Portoghese. Milan: Il Polifilo, 1978.

———. *Della trasportatione dell'obelisco vaticano et delle fabriche di nostro signore Papa Sisto V, fatte dal cavallier Domenico Fontana, architetto di Sua Santita.* Rome: Domenico Basa, 1590.

Fontana, Vincenzo. "Elementi per una biografia di M. Fabio Calvo Ravenneta." In *Vitruvio e Raffaello,* edited by Fontana and Morachiello, 45–61.

———. *Fra' Giovanni Giocondo Architetto 1433–c. 1515.* Vicenza: Neri Pozza Editore, 1988.

Fontana, Vincenzo, and Paolo Morachiello, eds. *Vitruvio e Raffaello: Il "De architectura" di Vitruvio nella traduzione inedita di Fabio Calvo Ravennate.* Rome: Officina Edizioni, 1975.

Francesco di Giorgio Martini. *La traduzione del "De architectura" di Vitruvio dal ms. II.I.141 della Biblioteca Nazionale Centrale di Firenze.* Edited by Marco Biffi. Pisa: Scuola Normale Superiore, 2002.

———. *Trattati di architettura ingegneria e arte militare.* 2 vols. Edited by Corrado Maltese. Transcription by Livia Maltese Degrassi. Milan: Edizioni Il Polifilo, 1967.

Franci, Raffaella. "Mathematics in the Manuscript of Michael of Rhodes." In *Book of Michael of Rhodes*, edited by Long, McGee, and Stahl, 3:115–146.

Frede, Michael. "Aristotle's Rationalism." In *Rationality in Greek Thought*, edited by Michael Frede and Gisela Striker, 157–173. New York: Oxford University Press, 1996.

French, Roger. *Ancient Natural History: Histories of Nature*. London: Routledge, 1994.

Freudenthal, Gideon. "Introductory Note." *Science in Context* 1 (1987): 105–108.

———. "Towards a Social History of Newtonian Mechanics: Boris Hessen and Henryk Grossmann Revisited." In *Scientific Knowledge Socialized*, edited by Imre Hronszky, Márta Fehér, and Balázs Dajka, 193–212. Dordrecht: Kluwer Academic, 1988.

Freudenthal, Gideon, and Peter McLaughlin. "Boris Hessen: In Lieu of a Biography." In *Social and Economic Roots*, edited by Freudenthal and McLaughlin, 253–256.

———. "Classical Marxist Historiography of Science: The Hessen-Grossmann Thesis." In *Social and Economic Roots*, edited by Freudenthal and McLaughlin, 1–38.

———, eds. *The Social and Economic Roots of the Scientific Revolution: Texts by Boris Hessen and Henryk Grossmann*. [Dordrecht]: Springer, 2009.

Gadol, Joan. *Leon Battista Alberti: Universal Man of the Early Renaissance*. Chicago: University of Chicago Press, 1969.

Galilei, Galileo. *Operations of the Geometric and Military Compass, 1606*. Translated by Stillman Drake. Washington, D.C.: Smithsonian Institution Press, 1978.

Galison, Peter. "Computer Simulations and the Trading Zone." In *The Disunity of Science*, edited by Galison and Stump, 118–157.

———. *Image and Logic: A Material Culture of Microphysics*. Chicago: University of Chicago Press, 1997.

———. "Trading with the Enemy." In *Trading Zones and Interactional Expertise*, edited by Michael E. Gorman, 25–52. Cambridge, Mass.: MIT Press, 2010.

Galison, Peter, and David J. Stump, eds. *The Disunity of Science: Boundaries, Contexts, and Power*. Stanford, Calif.: Stanford University Press, 1996.

Galluzzi, Paolo. "The Career of a Technologist." In *Leonardo da Vinci: Engineer*, edited by Galluzzi, 41–109.

———, ed. *Leonardo da Vinci: Engineer and Architect*. Montreal: Montreal Museum of Fine Arts, 1987.

———, ed. *Prima di Leonardo: Cultura delle macchine a Siena nel Rinascimento*. Milan: Electa, 1991.

Gambuti, Alessandro. "Nuove ricerche sugli 'Elementa Picturae.'" *Studi e Documenti di Architettura*, no. 1 (December 1972): 131–172.

Garin, Eugenio. "La letteratura degli umanisti." In *Storia della letteratura Italiana: Il quattrocento e l'Ariosto*, rev. ed., edited by Lucio Felici, 3:7–368. Milan: Garzanti, 1988.

Gatti Perer, Maria Luisa, and Alessandro Rovetta, eds. *Cesare Cesariano e il classicismo di primo Cinquecento.* Milan: Vita e Pensiero, 1996.

Gerbino, Anthony, and Stephen Johnston. *Compass and Rule: Architecture as Mathematical Practice in England.* Oxford: Museum of the History of Science, 2009.

Ghiberti, Lorenzo. *I commentarii (Biblioteca Nazionale Centrale di Firenze, II, I, 333).* Edited by Lorenzo Bartoli. Florence: Giunti Gruppo Editoriale, 1998.

Gieryn, Thomas F. "Distancing Science from Religion in Seventeenth-Century England." *Isis* 79 (December 1988): 582–593.

Giocondo, Giovanni, ed. *M. Vitruvius et Frontinus a Jocundo revisi repurgatique quantum ex collatione licuit.* Florence: Filippo Giunta, 1513.

———. *M. Vitruvius per Iocundum solito castigatior factus cum figures et tabula . . . .* Venice: Ioannis de Tridino alias Tacuino, 1511

Goldthwaite, Richard A. *The Building of Renaissance Florence: An Economic and Social History.* Baltimore: Johns Hopkins University Press, 1980.

———. *The Economy of Renaissance Florence.* Baltimore: Johns Hopkins University Press, 2009.

Golinski, Jan. *Making Natural Knowledge: Constructivism and the History of Science.* Cambridge: Cambridge University Press, 1998.

Goodman, David. *Power and Penury: Government, Technology, and Science in Philip II's Spain.* Cambridge: Cambridge University Press, 1988.

———. *Spanish Naval Power, 1589–1665: Reconstruction and Defeat.* Cambridge: Cambridge University Press, 1997.

Götz, Christian M., and Thomas Pankratz. "Edgar Zilsels Wirken im Rahmen der wiener Volksbildung und Lehrerfortbildung." In *Wien-Berlin-Prag*, edited by Haller and Stadler, 467–473.

Grafton, Anthony. *Leon Battista Alberti: Master Builder of the Italian Renaissance.* New York: Hill and Wang, 2000.

Graham, Loren R. "The Socio-political Roots of Boris Hessen: Soviet Marxism and the History of Science." *Social Studies of Science* 15 (November 1985): 705–722.

Grant, Edward. *The Foundations of Modern Science in the Middle Ages: Their Religious, Institutional, and Intellectual Contexts.* Cambridge: Cambridge University Press, 1996.

Grendler, Paul F. *The Universities of the Italian Renaissance.* Baltimore: Johns Hopkins University Press, 2002.

Grossmann, Henryk. "Die gesellschaftlichen Grundlagen der mechanistischen Philosophie und die Manufaktur." *Zeitschrift für Sozialforschung* 4, no. 2 (1935): 161–231.

———. "The Social Foundations of Mechanistic Philosophy and Manufacture." Translated by Gabriella Shalit. *Science in Context* 1 (March 1987): 109–180.

Gruber, Helmut. *Red Vienna: Experiment in Working-Class Culture, 1919–1934.* New York: Oxford University Press, 1991.

Gualdo, R. "Fabio Calvo, Marco." In *Dizionario biografico degli Italiani,* 43:723–727.

Guilmartin, John Francis, Jr. *Gunpowder and Galleys: Changing Technology and Mediterranean Warfare at Sea in the Sixteenth Century.* Cambridge: Cambridge University Press, 1974.

Günther, Hubertus. "Society in Filarete's *Libro architettonico* between Realism, Ideal, Science Fiction, and Utopia." In *Architettura e umanismo,* edited by Hub, 56–80.

Hacking, Ian. "The Disunities of the Sciences." In *The Disunity of Science,* edited by Galison and Stump, 37–74.

Hall, A. Rupert. *Ballistics in the Seventeenth Century: A Study in the Relations of Science and War with Reference Principally to England.* Cambridge: Cambridge University Press, 1952.

———. "Merton Revisited, or Science and Society in the Seventeenth Century." In *History of Science: An Annual Review of Literature, Research and Teaching,* vol. 2, edited by A. C. Crombie and M. A. Hoskin, 1-16. Cambridge: W. Heffer and Sons, 1963.

———. "The Scholar and the Craftsman in the Scientific Revolution." In *Critical Problems in the History of Science,* edited by Marshall Clagett, 3–23. Madison: University of Wisconsin Press, 1959.

Hall, Bert S. *Weapons and Warfare in Renaissance Europe.* Baltimore: Johns Hopkins University Press, 1997.

Haller, Rudolf, and Friedrich Stadler, eds. *Wien-Berlin-Prag: Der Aufstieg der wissenschaftlichen Philosophie.* Vienna: Verlag Hölder-Pichler-Tempsky, 1993.

Harcourt, Glenn. "Andreas Vesalius and the Anatomy of Antique Sculpture." *Representations* 17 (Winter 1987): 28–61.

Hardcastle, Garry L., and Alan W. Richardson. *Logical Empiricism in North America.* Minnesota Studies in the Philosophy of Science 18. Minneapolis: University of Minnesota Press, 2003.

Harkness, Deborah E. *The Jewel House: Elizabethan London and the Scientific Revolution.* New Haven, Conn.: Yale University Press, 2007.

Hart, Vaughan, and Peter Hicks. "On Sebastiano Serlio: Decorum and the Art of Architectural Invention." In *Paper Palaces,* edited by Hart and Hicks, 40–157.

———. *Palladio's Rome: A Translation of Andrea Palladio's Two Guidebooks to Rome.* New Haven, Conn.: Yale University Press, 2006.

———, eds. *Paper Palaces: The Rise of the Renaissance Architectural Treatise.* New Haven, Conn.: Yale University Press, 1998.

Heintel, Peter. *System und Ideologie: Der Austromarxismus im Spiegel der Philosophie Max Adlers.* Vienna: Oldenbourg, 1967.

Henninger-Voss, Mary J. "Comets and Cannonballs: Reading Technology in a Sixteenth-Century Library." In *The Minajul Hand,* edited by Roberts, Schaffer, and Dear, 10–31.

———. "How the 'New Science' of Cannons Shook Up the Aristotelian Cosmos." *Journal of the History of Ideas* 63 (July 2002): 371–397.

Henry, John. *The Scientific Revolution and the Origins of Modern Science.* 3d ed. Basingstoke, Eng.: Palgrave Macmillan, 2008.

Hessen, Boris. "The Social and Economic Roots of Newton's 'Principia.'" In *Science at the Crossroads,* 2d ed., introduction by P. G. Werskey, 149–212. London: Frank Cass, 1971.

———. "The Social and Economic Roots of Newton's *Principia.*" In *Social and Economic Roots,* edited by Freudenthal and McLaughlin, 41–101.

Howard, Don. "Two Left Turns Make a Right: On the Curious Political Career of North American Philosophy of Science at Midcentury." In *Logical Empiricism in North America,* edited by Hardcastle and Richardson, 25–93.

Hoyningen-Huene, Paul. *Reconstructing Scientific Revolutions: Thomas S. Kuhn's Philosophy of Science.* Translated by Alexander T. Levine. Chicago: University of Chicago Press, 1993.

Hub, Berthold, ed. *Architettura e umanismo: Nuovi studi su Filarete. Arte Lombarda,* n.s. 155, 1 (2009).

Iliffe, Rob. "Material Doubts: Hooke, Artisan Culture, and the Exchange of Information in 1670s London." *British Journal for the History of Science* 28, no. 3 (1995): 285–318.

Jacks, Philip J. "Calvo, Marco Fabio." In *Dictionary of Art,* edited by Turner, 5:448.

Janson, H. W. "Titian's Laocoon Caricature and the Vesalian-Galenist Controversy." *Art Bulletin* 28 (March 1946): 49–53.

Jardine, Lisa. *Worldly Goods: A New History of the Renaissance.* New York: Doubleday, 1996.

Jardine, Nicholas. "Essay Review: Zilsel's Dilemma." *Annals of Science* 60, no. 1 (2003): 85–94.

Jay, Martin. *The Dialectical Imagination: A History of the Frankfurt School and the Institute of Social Research, 1923–1950.* London: Heinemann, 1973.

———. *Marxism and Totality: The Adventures of a Concept from Lukács to Habermas.* Berkeley: University of California Press, 1984.

———. *Permanent Exiles: Essays on the Intellectual Migration from Germany to America.* New York: Columbia University Press, 1986.

Jobé, Joseph, ed. *Guns: An Illustrated History of Artillery.* Greenwich, Conn.: New York Graphic Society, 1971.

Johns, Adrian. "Coffeehouses and Print Shops." In *The Cambridge History of Science,* vol. 3: *Early Modern Science,* edited by Park and Daston, 320–340.

———. *The Nature of the Book: Print and Knowledge in the Making.* Chicago: University of Chicago Press, 1998.

Johnston, Steven. "Making Mathematical Practice: Gentlemen, Practitioners, and Artisans in Elizabethan England." Ph.D. diss., Cambridge University, 1994. http://www.mhs.ox.ac.uk/staff/saj/thesis/abstract.htm.

Jones, Barry, Andrea Sereni, and Massimo Ricci. "Building Brunelleschi's Dome: A Practical Methodology Verified by Experiment." *Journal of the Society of Architectural Historians* 69 (March 2010): 39–61.

Jones, William David. "Toward a Theory of Totalitarianism: Franz Borkenau's *Pareto.*" *Journal of the History of Ideas* 53 (July–September 1992): 455–466.

Karmon, David. "Restoring the Ancient Water Supply System in Renaissance Rome: The Popes, the Civic Administration, and the Acqua Vergine." In *Aqua urbis Romae: The Waters of the City of Rome.* http://www.iath.virginia.edu/waters.

Kaufmann, Thomas DaCosta. *Arcimboldo: Visual Jokes, Natural History, and Still-Life Painting.* Chicago: University of Chicago Press, 2009.

Kellenbenz, Hermann. *The Rise of the European Economy: An Economic History of Continental Europe from the Fifteenth to the Eighteenth Century.* Edited by Gerhard Benecke. London: Weidenfeld and Nicolson, 1976.

Kemp, Martin. "A Drawing for the *Fabrica*; and Some Thoughts upon the Vesalius Muscle-Men." *Medical History* 14 (July 1970): 277–288.

———. Introduction to Alberti, *On Painting.* Introduction and notes by Kemp, translated by Grayson. London: Penguin Books, 1991.

———. *Leonardo da Vinci: The Marvellous Works of Nature and Man.* Rev. ed. Oxford: Oxford University Press, 2006.

———. "'The Mark of Truth': Looking and Learning in Some Anatomical Illustrations from the Renaissance and Eighteenth Century." In *Medicine and the Five Senses*, edited by W. F. Bynum and Roy Porter, 85–121. Cambridge: Cambridge University Press, 1993.

———. *The Science of Art: Optical Themes in Western Art from Brunelleschi to Seurat.* New Haven, Conn.: Yale University Press, 1990.

———. "Wrought by No Artist's Hand: The Natural, the Artificial, the Exotic, and the Scientific in Some Artifacts from the Renaissance." In *Reframing the Renaissance: Visual Culture in Europe and Latin America, 1450–1650*, edited by Claire Farago, 177–196. New Haven, Conn.: Yale University Press, 1995.

Kibre, Pearl, and Nancy G. Siraisi. "The Institutional Setting: The Universities." In *Science in the Middle Ages*, edited by David C. Lindberg, 120–144. Chicago: University of Chicago Press, 1978.

King, Ross. *Brunelleschi's Dome: How a Renaissance Genius Reinvented Architecture.* New York: Penguin Books, 2001.

Klein, Ursula. "Essay Review: Styles of Experimentation and Alchemical Matter Theory in the Scientific Revolution." *Metascience* 16, no. 2 (2007): 247–256.

Kluke, Paul. *Die Stiftungsuniversität Frankfurt am Main, 1914–1932.* Frankfurt am Main: Verlag von Waldemar Kramer, 1972.

Kovaly, Pavel. "Arnošt Kolman: Portrait of a Marxist-Leninist Philosopher." *Studies in Soviet Thought* 12 (December 1972): 337–366.

Koyré, Alexandre. *Études galiléennes.* Paris: Hermann, 1966.

———. *From the Closed World to the Infinite Universe.* Baltimore: Johns Hopkins University Press, 1957.

Kraschewski, Hans-Joachim. *Wirtschaftspolitik im deutschen Territorialstaat des 16. Jahrhunderts: Herzog Julius von Braunschweig-Wolfenbüttel (1528–1589).* Cologne: Böhlau Verlag, 1978.

Krautheimer, Richard, in collaboration with Trude Krautheimer-Hess. *Lorenzo Ghiberti.* 3d ed. Princeton, N.J.: Princeton University Press, 1982.

Kraye, Jill, ed. *The Cambridge Companion to Renaissance Humanism.* Cambridge: Cambridge University Press, 1996.

Krinsky, Carol H. "Seventy-Eight Vitruvius Manuscripts." *Journal of the Warburg and Courtauld Institutes* 30 (1967): 36–70.

Kristeller, Paul Oskar. "The Modern System of the Arts." In Kristeller, *Renaissance Thought. II: Papers on Humanism and the Arts,* 163–227. New York: Harper and Row, Harper Torchbooks, 1965.

Krohn, Wolfgang, and Diederick Raven. "The 'Zilsel Thesis' in the Context of Edgar Zilsel's Research Programme." *Social Studies of Science* 30 (December 2000): 925–933.

Kuhn, Rick. *Henryk Grossman and the Recovery of Marxism.* Urbana: University of Illinois Press, 2007.

Kuhn, Thomas S. *The Structure of Scientific Revolutions.* 2d ed. Chicago: University of Chicago Press, 1970.

Lamprey, John P. "An Examination of Two Groups of Georg Hartmann Sixteenth-century Astrolaes and the Tables Used in Their Manufacture." *Annals of Science* 54, no. 2 (1997): 111–142.

Lane, Frederic Chapin. *Venetian Ships and Shipbuilders of the Renaissance.* Baltimore: Johns Hopkins University Press, 1934.

Lang, S. "Sforzinda, Filarete and Filelfo." *Journal of the Warburg and Courtauld Institutes* 35 (1972): 391–397.

Lefaivre, Liane. *Leon Battista Alberti's Hypnerotomachia Poliphili: Re-Cognizing the Architectural Body in the Early Italian Renaissance.* Cambridge, Mass.: MIT Press, 1997.

Leng, Rainer. *Ars belli: Deutsche taktische und kriegstechnische Bilderhandschriften und Traktate im 15. und 16. Jahrhundert.* 2 vols. Wiesbaden: Reichert Verlag, 2002.

Leonardo da Vinci. *The Madrid Codices.* 5 vols. Edited and translated by Ladislao Reti. New York: McGraw-Hill, 1974.

Leser, Norbert. *Zwischen Reformismus und Bolschewismus: Der Austromarxismus als Theorie und Praxis.* Vienna: Europa Verlag, 1968.

Lewis, Douglas. "Maser, Villa Barbaro." In *Dictionary of Art,* edited by Turner, 20:545–547.

Lindberg, David C. "Experiment and Experimental Science." In *The Oxford Dictionary of the Middle Ages,* 4 vols., edited by Robert E. Bjork, 2:604–605. Oxford: Oxford University Press, 2010.

Lloyd, G. E. R. "Experiment in Early Greek Philosophy and Medicine." In Lloyd, *Methods and Problems of Greek Science: Selected Papers,* 70–99. Cambridge: Cambridge University Press, 1991.

————. *Magic, Reason, and Experience: Studies in the Origin and Development of Greek Science*. Cambridge: Cambridge University Press, 1979.

Long, Pamela O. "The Contribution of Architectural Writers to a 'Scientific' Outlook in the Fifteenth and Sixteenth Centuries." *Journal of Medieval and Renaissance Studies* 15 (Fall 1985): 265–298.

————. "Hydraulic Engineering and the Study of Antiquity: Rome, 1557–70." *Renaissance Quarterly* 61 (Winter 2008): 1098–1138.

————. "Introduction: The World of Michael of Rhodes, Venetian Mariner." In *Book of Michael of Rhodes*, edited by Long, McGee, and Stahl, 3:1–33.

————. "Objects of Art/Objects of Nature: Visual Representation and the Representation of Nature." In *Merchants and Marvels*, edited by Smith and Findlen, 63–82.

————. "The Openness of Knowledge: An Ideal and Its Context in 16th-Century Writings on Mining and Metallurgy." *Technology and Culture* 32 (April 1991): 318–355.

————. *Openness, Secrecy, Authorship: Technical Arts and the Culture of Knowledge from Antiquity to the Renaissance*. Baltimore: Johns Hopkins University Press, 2001.

————. "Picturing the Machine: Francesco di Giorgio and Leonardo da Vinci in the 1490s." In *Picturing Machines, 1400–1650*, edited by Wolfgang Lefèvre, 117–141. Cambridge, Mass.: MIT Press, 2004.

————. "Plants and Animals in History: The Study of Nature in Renaissance and Early Modern Europe." *Historical Studies in the Natural Sciences* 38 (Spring 2008): 313–323.

Long, Pamela O., David McGee, and Alan M. Stahl, eds. *The Book of Michael of Rhodes: A Fifteenth-Century Maritime Manuscript*. 3 vols. Cambridge, Mass.: MIT Press, 2009.

Lopez, Robert S. *The Commercial Revolution of the Middle Ages, 950–1350*. Cambridge: Cambridge University Press, 1976.

Lowenthal, Richard. "In Memoriam Franz Borkenau." *Der Monat* 9 (July 1957): 57–60.

Lukács, Georg. *History and Class Consciousness: Studies in Marxist Dialectics*. Translated by Rodney Livingstone. Cambridge, Mass.: MIT Press, [1971].

Maffioli, Cesare S. *La via delle acque (1500–1700): Appropriazione delle arti e trasformazione delle matematiche*. Florence: Leo S. Olschki, 2010.

Maier, Jessica. "Mapping Past and Present: Leonardo Bufalini's Plan of Rome (1551)." *Imago Mundi* 59, no. 1 (2007): 1–23.

Manetti, Antonio di Tuccio. *The Life of Brunelleschi*. Edited by Howard Saalman. Translated by Catherine Enggass. University Park: Pennsylvania State University Press, 1970.

Marchis, Vittorio. "Nuove dimensioni per l'energia: le macchine di Francesco di Giorgio." In *Prima di Leonardo*, edited by Galluzzi, 113–120.

Marcucci, Laura. "Giovanni Sulpicio e la prima edizione del *De architectura* di Vitruvio." *Studi e Documenti di Architettura*, no. 8 (September 1978): 185–195.

Marx, Karl. *Capital: A Critique of Political Economy.* Translated by Ben Fowkes. Vol. 1. London: Penguin Books, 1976.

Masini, Francesco. *Discorso di Francesco Masini sopra un modo nuovo, facile, e reale, di trasportar su la Piazza di San Pietro la guglia, ch'è in Roma, detta di Cesare.* Cesena: Bartolomeo Raverij, 1586.

Mauss, Marcel. *The Gift: Form and Reason for Exchange in Archaic Societies.* Translated by W. D. Halls. New York: W. W. Norton, 1990.

Mayer, Anna-K. "Fatal Mutilations: Educationism and the British Background to the 1931 International Congress for the History of Science and Technology." *History of Science* 40 (December 2002): 445–472.

———. "Setting Up a Discipline, II: British History of Science and 'The End of Ideology,' 1931–1948." *Studies in the History and Philosophy of Science* 35 (March 2004): 41–72.

Mazzotti, Massimo, ed. *Knowledge as Social Order: Rethinking the Sociology of Barry Barnes.* Aldershot, Eng.: Ashgate, 2008.

McEwen, Indra Kagis. *Vitruvius: Writing the Body of Architecture.* Cambridge, Mass.: MIT Press, 2003.

McGee, David. "The Shipbuilding Text of Michael of Rhodes." In *Book of Michael of Rhodes*, edited by Long, McGee, and Stahl, 3:211–241.

Meijer, Bert W. "Calcar, Jan Steven [Johannes Stephanus] van." In *Dictionary of Art,* edited by Turner, 5:415–416.

Mercati, Michele. *De gli obelischi di Roma.* Rome: Domenico Basa, 1589.

———. *Gli obelischi di Roma.* Edited by Gianfranco Cantelli. Bologna: Cappelli Editore, 1981.

Merton, Robert K. *Science, Technology and Society in Seventeenth-Century England.* 1938. [Atlantic Highlands], N.J.: Humanities Press, 1978.

Meserve, Margaret. "Nestor Denied: Francesco Filelfo's Advice to Princes on the Crusade against the Turks." In *Expertise: Practical Knowledge and the Early Modern State,* edited by Eric H. Ash. *Osiris,* 2d ser., 25 (2010): 47–65.

Moran, Bruce T. "Courts and Academies." In *The Cambridge History of Science,* vol. 3: *Early Modern Science,* edited by Park and Daston, 251–271.

———, ed. *Patronage and Institutions: Science, Technology, and Medicine at the European Court, 1500–1750.* Rochester, N.Y.: Boydell Press, 1991.

Munck, Bert de. "Corpses, Live Models, and Nature: Assessing Skills and Knowledge before the Industrial Revolution (Case: Antwerp)." *Technology and Culture* 51 (April 2010): 332–356.

Muraro, Michelangelo. "Tiziano e le anatomie del Vesalio." In *Tiziano e Venezia: Convegno Internazionale di Studi, Venezia, 1976,* 307–316. Vicenza: Neri Pozza, 1980.

Mussini, Massimo. *Francesco di Giorgio e Vitruvio: Le traduzioni del "De architectura" nei codici Zichy, Spencer 129 e Magliabechiano II.I.141.* 2 vols. Mantua: Fondazione Centro Studi L. B. Alberti, 2002.

———. *Il Trattato di Francesco di Giorgio Martini e Leonardo: Il Codice Estense restituito.* Parma: Università di Parma, 1991.

———. "La trattatistica di Francesco di Giorgio: un problema critico aperto." In *Francesco di Giorgio, architetto,* edited by Fiore and Tafuri, 358–379.

———. "Un frammento del *Trattato* di Francesco di Giorgio Martini nell'archivio di G. Venturi alla Biblioteca Municipale di Reggio Emilia." In *Prima di Leonardo,* edited by Galluzzi. Milan: Electa, 1991.

Nef, John U. "Mining and Metallurgy in Medieval Civilisation." In *The Cambridge Economic History of Europe,* 2d ed., vol. 2: *Trade and Industry in the Middle Ages,* edited by M. M. Postan and Edward Miller, assisted by Cynthia Postan, 691–761, 933–940. Cambridge: Cambridge University Press, 1987.

Neurath, Otto. *Empiricism and Sociology.* Edited by Marie Neurath and Robert S. Cohen. Translated by Paul Foulkes and Marie Neurath. Dordrecht: D. Reidel, 1973.

Newman, William R. "Alchemical Atoms or Artisanal 'Building Blocks'? A Response to Klein." *Perspectives on Science* 17, no. 2 (2009): 212–231.

———. "Brian Vickers on Alchemy and the Occult: A Response." *Perspectives on Science* 17, no. 4 (2009): 482–506.

———. *Promethean Ambitions: Alchemy and the Quest to Perfect Nature.* Chicago: University of Chicago Press, 2004.

———. "Technology and Alchemical Debate in the Late Middle Ages." *Isis* 80 (September 1989): 423–445.

Newman, William R., and Lawrence M. Principe. *Alchemy Tried in the Fire: Starkey, Boyle, and the Fate of Helmontian Chymistry.* Chicago: University of Chicago Press, 2002.

Nicholas of Cusa. *Idiota de Mente: The Layman: About Mind.* Translated by Clyde Lee Miller. New York: Abaris, 1979.

———. *The Idiot in Four Books.* Anonymous translator. London: William Leake, 1650.

———. *Opera omnia.* Vol. 5: *Idiota: De sapientia, De mente, De staticis experimentis.* Rev. ed. Edited by Renata Steiger et al. from the edition of Ludwig Baur. Hamburg: Felix Meiner, 1983.

Nummedal, Tara. *Alchemy and Authority in the Holy Roman Empire.* Chicago: University of Chicago Press, 2007.

Nutton, Vivian. "Introduction." Vesalius, *De humanis corporis fabrica,* edited by Daniel Garrison and Malcolm Hast. http://vesalius.northwestern.edu.

Nye, Mary Jo. "Re-Reading Bernal: History of Science at the Crossroads in 20th-Century Britain." In *Aurora Torealis: Studies in the History of Science and Ideas in Honor of Tore Fränsmyr,* edited by Marco Beretta, Karl Grandin, and Svante Lindquist, 235–258. Sagamore Beach, Mass.: Science History Publications, 2008.

Ogilvie, Brian W. *The Science of Describing: Natural History in Renaissance Europe* (Chicago: University of Chicago Press, 2006).

Olschki, Leonardo. *Geschichte der neusprachlichen wissenschaftlichen Literatur:* Vol. 1: *Die Literatur der Technik und der angewandten Wissenschaften von Mittelalter bis zur Renaissance.* Heidelberg: Winter, 1919. Vol. 2: *Bildung und*

Wissenschaft im Zeitalter der Renaissance in Italien. Leipzig: Olschki, 1922. Vol. 3: *Galilei und seine Zeit.* Halle: Niemeyer, 1927.

O'Malley, C. D. *Andreas Vesalius of Brussels, 1514–1564.* Berkeley: University of California Press, 1964.

Onians, John. "Alberti and Filarete: A Study in Their Sources." *Journal of the Warburg and Courtauld Institutes* 34 (1971): 96–114.

Ovitt, George, Jr. *The Restoration of Perfection: Labor and Technology in Medieval Culture.* New Brunswick, N.J.: Rutgers University Press, 1987.

Owens, Joseph. "The Universality of the Sensible in the Aristotelian Noetic." In *Aristotle: The Collected Papers of Joseph Owens,* edited by John R. Catan, 59–73. Albany: State University of New York Press, 1981.

Pagliara, P. N. "Giovanni Giocondo da Verona (Fra Giocondo)." In *Dizionario biografico degli Italiani,* 56:326–338.

Palladio, Andrea. *The Four Books on Architecture.* Translated by Robert Tavernor and Richard Schofield. Cambridge, Mass.: MIT Press, 1997.

———. *I quattro libri dell'architettura di Andrea Palladio.* Venice: Dominico de'Franceschi, 1570.

Park, Katharine. *Secrets of Women: Gender, Generation, and the Origins of Human Dissection.* New York: Zone Books, 2006.

Park, Katharine, and Lorraine Daston, eds. *The Cambridge History of Science.* Vol. 3: *Early Modern Science.* Cambridge: Cambridge University Press, 2006.

Pasquale, Salvatore Di. "Leonardo, Brunelleschi, and the Machinery of the Construction Site." In *Leonardo da Vinci: Engineer and Architect,* edited by Galluzzi, 163–181.

Payne, Alina A. *The Architectural Treatise in the Italian Renaissance: Architectural Invention, Ornament, and Literary Culture.* Cambridge: Cambridge University Press, 1999.

Pérez-Gómez, Alberto. "The *Hypnerotomachia Poliphili* by Francesco Colonna: The Erotic Nature of Architectural Meaning." In *Paper Palaces,* edited by Hart with Hicks, 86–104.

Pérez-Ramos, Antonio. *Francis Bacon's Idea of Science and the Maker's Knowledge Tradition.* Oxford: Clarendon, 1988.

Peto, Luca. *De mensuris et ponderibus Romanis et Graecis.* Venice: [P. Manutius], 1573.

———. *Discorso di Luca Peto intorno alla cagione della Eccessiva Inondatione del Tevere in Roma, et modo in parte di soccorrervi.* Rome: Giuseppe degl'Angeli, 1573.

Petrović, Gajo. "Reification." In *Dictionary of Marxist Thought,* edited by Bottomore, 411–413.

Pfabigan, Alfred. *Max Adler: Eine politische Biographie.* Frankfurt: Campus, 1982.

Phillips, Carla Rahn. *Six Galleons for the King of Spain: Imperial Defense in the Early Seventeenth Century.* Baltimore: Johns Hopkins University Press, 1986.

Piccolpasso, Cipriano. *Le piante et i ritratti delle città e terre dell' Umbria sottoposte al governo di Perugia.* Edited by Giovanni Cecchini. Rome: Istituto Nazionale d'Archeologia e Storia dell'Arte, 1963.

———. *The Three Books of the Potter's Art: A Facsimile of the Manuscript in the Victoria and Albert Museum, London.* 2 vols. Translated and introduced by Ronald Lightbown and Alan Caiger-Smith. London: Scholar Press, 1980.

Pigafetta, Filippo. *Discorso di M. Filippo Pigafetta d'intorno all'historia della Aguglia, et alla ragione del muoverla.* Rome: Bartolomeo Graffi, 1586.

Pigeaud, Jackie "Formes et normes dans le 'De fabrica' de Vésale." In *Le corps à la Renaissance, Actes du XXXe Colloque de Tours, 1987,* edited by Jean Céard, Marie Madeleine Fontaine, and Jean-Claude Margolin, 399–421. Paris: Aux Amateurs de Livres, 1990.

Piovan, F. "Fausto, Vittore." *Dizionario biografico degli Italiani,* 45:398–401.

Plommer, Hugh. *Vitruvius and Later Roman Building Manuals.* London: Cambridge University Press, 1973.

Pollak, Martha D. *Military Architecture, Cartography, and the Representation of the Early Modern European City: A Checklist of Treatises on Fortification in the Newberry Library.* Chicago: Newberry Library, 1991.

Portuondo, Marìa M. *Secret Science: Spanish Cosmography and the New World.* Chicago: University of Chicago Press, 2009.

Prager, Frank D., and Gustina Scaglia. *Brunelleschi: Studies of His Technology and Inventions.* Cambridge, Mass.: MIT Press, 1970.

Principe, Lawrence M. *The Aspiring Adept: Robert Boyle and His Alchemical Quest.* Princeton, N.J.: Princeton University Press, 1998.

Propositiones Aristoteles. Venice: [Georgius Arrivabenus, ca. 1490].

Rabil, Albert, Jr., ed. *Renaissance Humanism: Foundations, Forms, and Legacy.* 3 vols. Philadelphia: University of Pennsylvania Press, 1988.

Rabinbach, Anson. *The Crisis of Austrian Socialism: From Red Vienna to Civil War, 1927–1934.* Chicago: University of Chicago Press, 1983.

Radke, Gary M., ed. *The Gates of Paradise: Lorenzo Ghiberti's Renaissance Masterpiece.* Atlanta: High Art Museum; New Haven, Conn.: Yale University Press, 2007.

Raven, Diederick. "Edgar Zilsel in America." In *Logical Empiricism in North America,* edited by Hardcastle and Richardson, 129–148.

———. "Edgar Zilsel's Research Programme: Unity of Science as an Empirical Problem." In *Vienna Circle and Logical Empiricism,* edited by Stadler, 225–234.

Raven, Diederick, and Wolfgang Krohn. "Edgar Zilsel: His Life and Work (1891–1944)." In Zilsel, *Social Origins of Modern Science,* edited by Raven, Krohn, and Cohen, xix–lix.

Rée, Jonathan. *Proletarian Philosophers: Problems in Socialist Culture in Britain, 1900–1940.* Oxford: Clarendon Press, 1984.

Reisch, George A. "Planning Science: Otto Neurath and the *International Encyclopedia of Unified Science.*" *British Journal for the History of Science* 27 (June 1994): 153–175.

Renn, Jürgen, and Matteo Valleriani. "Galileo and the Challenge of the Arsenal." *Nuncius* 16, no. 2 (2001): 481–503.

Rinne, Katherine Wentworth. *The Waters of Rome: Aqueducts, Fountains, and the Birth of the Baroque City.* New Haven, Conn.: Yale University Press, 2010.

Roberts, Lissa. Introduction to *The Mindful Hand*, edited by Roberts, Schaffer, and Dear, 1–8.

Roberts, Lissa, and Simon Schaffer. Preface to *The Mindful Hand*, edited by Roberts, Schaffer, and Dear, xiii–xxvii.

Roberts, Lissa, Simon Schaffer, and Peter Dear, eds. *The Mindful Hand: Inquiry and Invention from the Late Renaissance to Early Industrialisation.* Amsterdam: Koninklijke Nederlandse Akademie van Wetenschappen, 2007.

Romano, Antonella, ed. *Rome et la science moderne entre Renaissance et Lumières.* Rome: École Française de Rome, 2008.

Rosand, David, and Michelangelo Muraro. *Titian and the Venetian Woodcut.* Washington, D.C.: International Exhibitions Foundation, 1976–1977.

Rosenfeld, Myra Nan. "From Bologna to Venice and Paris: The Evolution and Publication of Sebastiano Serlio's Books I and II, *On Geometry* and *On Perspective*, for Architects." In *The Treatise on Perspective: Published and Unpublished*, edited by Lyle Massey, 280–321. Washington, D.C.: National Gallery of Art, 2003.

Ross, Sydney. "Scientist: The Story of a Word." *Annals of Science* 18 (June 1962): 65–85.

Rossi, Franco. "L'Arsenale: I quadri direttivi." In *Storia di Venezia: Dalle origini alla caduta della Serenissima*, vol. 5: *Il Rinascimento: Società ed economia*, edited by Alberto Tenenti and Ugo Tucci, 593–639. Rome: Istituto della Enciclopedia Italiana, 1996.

Rossi, Paolo. *Philosophy, Technology, and the Arts in the Early Modern Era.* Translated by Salvator Attanasio. Edited by Benjamin Nelson. New York: Harper and Row, 1970.

Rovetta, Alessandro, Elio Monducci, and Corrado Caselli. *Cesare Cesariano e il Rinascimento a Reggio Emilia.* Milan: Silvana, 2008.

Rowland, Ingrid D. *Culture of the High Renaissance: Ancients and Moderns in Sixteenth-Century Rome.* Cambridge: Cambridge University Press, 1998.

———. Introduction to *Ten Books on Architecture: The Corsini Incunabulum*, by Vitruvius, edited by Rowland, 1–31.

Russo, Valerie E. "Henryk Grossmann and Franz Borkenau: A Bio-Bibliography." *Science in Context* 1 (March 1987): 181–191.

———. "Profilo di Franz Borkenau." *Rivista di Filosofia* 72 (June 1981): 291–316.

Saalman, Howard. *Filippo Brunelleschi: The Cupola of Santa Maria del Fiore.* London: Zwemmer, 1980.

———. Introduction to *The Life of Brunelleschi* by Antonio di Tuccio Manetti, edited by Saalman and translated by Enggass.

Santing, Catrien. "Andreas Vesalius's *De Fabrica corporis humana*, Depiction of the Human Model in Word and Image." In *Body and Embodiment in Netherlandish Art*, Netherlands Yearbook for History of Art, 2007–2008, vol. 58, edited by Ann-Sophie Lehmann and Herman Roodenburg, 59–85. Zwolle: Waanders, 2008.

Scaglia, Gustina. *Francesco di Giorgio: Checklist and History of Manuscripts and Drawings in Autographs and Copies from ca. 1470 to 1687 and Renewed Copies (1764–1839)*. Bethlehem, Penn.: Lehigh University Press; and Cranbury, N.J.: Associated University Presses, 1992.

Schiefsky, Mark J. "Art and Nature in Ancient Mechanics." In *The Artificial and the Natural*, edited by Bensaude-Vincent and Newman, 67–108.

Schmidtchen, Volker. *Bombarden, Befestigungen, Büchsenmeister: Von den ersten Mauerbrechern des Spätmittelalters zur Belagerungsartillerie der Renaissance*. Düsseldorf: Droste, 1977.

Schmitt, Charles B. *Aristotle and the Renaissance*. Cambridge: Mass.: Harvard University Press, 1983.

Serlio, Sebastiano. *On Architecture*. Vol. 1: *Books I–V of "Tutte l'opere d'architettura et prospetiva."* Vol. 2: *Books VI and VII of "Tutte L'opere d'architettura et prospetiva" with "Castrametation of the Romans" and "The Extraordinary Book of Doors" by Sebastiano Serlio*. Translated with Introduction and Commentary by Vaughan Hart and Peter Hicks. New Haven, Conn.: Yale University Press, 1996, 2001.

———. *Regole generali di architectura sopra le cinque maniere de gliedifici, cioe, thoscano, dorico, ionico, corinthio, et composito, con gliessempi dell'antiquita, che per la magior parte concordano con la dottrina di Vitruvio*. Venice: F. Marcolini da Forli, 1537.

Sgarbi, Claudio, ed. *Vitruvio ferrarese "De architectura": La prima versione illustrate*. Modena: Franco Cosimo Panini, 2004.

Shapin, Steven. *The Scientific Revolution*. Chicago: University of Chicago Press, 1996.

———. "Understanding the Merton Thesis." *Isis* 79 (December 1988): 594–604.

Shapin, Steven, and Simon Schaffer. *Leviathan and the Air-Pump: Hobbes, Boyle, and the Experimental Life*. Princeton, N.J.: Princeton University Press, 1985.

Simons, Patricia, and Monique Kornell. "Annibal Caro's After-Dinner Speech (1536) and the Question of Titian as Vesalius's Illustrator." *Renaissance Quarterly* 61 (Winter 2008): 1069–1097.

Sinisgalli, Rocco. *Il nuovo "De Pictore" di Leon Battista Alberti / The New "De pictore" of Leon Battista Alberti*. Rome: Edizioni Kappa, 2006.

Siraisi, Nancy G. "Vesalius and Human Diversity in *De humani corporis fabrica*." *Journal of the Warburg and Courtauld Institutes* 57 (1994): 60–88.

———. "Vesalius and the Reading of Galen's Teleology." *Renaissance Quarterly* 50 (Spring 1997): 1–37.

Smith, Christine. *Architecture in the Culture of Early Humanism: Ethics, Aesthetics, and Eloquence, 1400–1470*. New York: Oxford University Press, 1992.

Smith, Pamela H. *The Body of the Artisan: Art and Experience in the Scientific Revolution*. Chicago: University of Chicago Press, 2004.

———. "In a Sixteenth-Century Goldsmith's Workshop." In *The Mindful Hand*, edited by Roberts, Schaffer, and Dear, 32–57.

Smith, Pamela H., and Tonny Beentjes. "Nature and Art, Making and Knowing: Reconstructing Sixteenth-Century Life-Casting Techniques." *Renaissance Quarterly* 63 (Spring 2010): 128–179.

Smith, Pamela H., and Paula Findlen, eds. *Merchants and Marvels: Commerce, Science, and Art in Early Modern Europe*. New York: Routledge, 2002.

Soubiran, Jean, ed. and trans. *Vitruve De L'architecture, Livre IX*. Paris: Société D'Édition "Les Belle Lettres," 1969.

Spies, Gerd, ed. *Technik der Steingewinnung und der Flussschiffahrt in Harzvoland in früher Neuzeit*. Braunschweig: Waisenhaus, 1992.

Stadler, Friedrich. "Aspects of the Social Background and Position of the Vienna Circle at the University of Vienna." In *Rediscovering the Forgotten Vienna Circle*, edited by Uebel, 51–77.

———. "What Is the Vienna Circle? Some Methodological and Historiographical Answers." In *The Vienna Circle and Logical Empiricism*, edited by Stadler, xi–xxiii.

———, ed. *The Vienna Circle and Logical Empiricism: Re-evaluation and Future Perspectives*. Dordrecht: Kluwer Academic, 2003.

Stahl, Alan M. "Michael of Rhodes: Mariner in Service to Venice." In *The Book of Michael of Rhodes*, edited by Long, McGee, and Stahl, 3:35–98.

Stewart, Richard W. *The English Ordnance Office, 1585–1625: A Case Study in Bureaucracy*. Woodbridge, Suffolk: Boydel Press, 1996.

Sulpicius, Giovanni, ed. *L. Vitruvii Pollionis ad Cesarem Augustum De Architectura Liber Primus (–Decimus)*. Rome: [Giorgio Herolt or Eucarius Silber], [1486–1492?].

Tafuri, Manfredo. "Cesare Cesariano e gli studi Vitruviani nel Quattrocento." In *Scritti rinascimentali di architettura*, edited by Analdo Bruschi, Corrado Maltese, Manfredo Tafuri, and Renato Bonelli. Milan: Il Polifilo, 1978.

———. "Daniele Barbaro e la cultura scientifica veneziana del'500." In *Cultura, scienze e technica nella Venezia del cinquecento: Giovan Battista Benedetti e il suo tempo*, 55–81. Venice: Istituto Veneto di Scienze, Lettere ed Arti, 1987.

Tartaglia, Niccolò. *Nova scientia inventa da Nicolo Tartalea B*. Venice: Stephano da Sabio, 1537.

———. *Quesiti et inventioni diverse de Nicolo Tartalea Brisciano*. Venice: Venturino Ruffinelli for N. Tartaglia, 1546.

Tashjean, John E. "Borkenau: The Rediscovery of a Thinker." *Partisan Review* 51, no. 2 (1984): 289–300.

———. "Franz Borkenau: A Study of His Social and Political Ideas." Ph.D. diss., Georgetown University, 1962.

Teesdale, Edmund B. *Gunfounding in the Weald in the Sixteenth Century.* London: Trustees of the Royal Armouries, 1991.

———. *The Queen's Gunstonemaker: Being an Account of Ralph Hogge, Elizabethan Ironmaster and Gunfounder.* Seaford, Eng.: Lindel, 1984.

Thoren, Victor E. *The Lord of Uraniborg: A Biography of Tycho Brahe.* Cambridge: Cambridge University Press, 1990.

Toledano, Ralph. *Francesco di Giorgio Martini: Pittore e scultore.* Milan: Electa, 1987.

Trevisi, Antonio. *Fondamento del edifitio nel quale si tratta con la santita de N.S. Pio Papa IIII sopra la innondatione del fiume.* Rome: Antonio Blado, 1560.

Turner, Jane, ed. *The Dictionary of Art.* 34 vols. New York: Grove Dictionaries, 1996.

Uebel, Thomas E., ed. *Rediscovering the Forgotten Vienna Circle: Austrian Studies on Otto Neurath and the Vienna Circle.* Dordrecht: Kluwer Academic, 1991.

———. "Vienna Circle." *Stanford Encyclopedia of Philosophy.* http://plato. stanford.edu/entries/Vienna-circle/. June 28, 2006. Revised 18 September 2006.

Vagnetti, Luigi. "Considerazioni sui Ludi Matematici." *Studi e Documenti di Architettura*, no. 1 (December 1972): 173–259.

Vagnetti, Luigi, and Laura Marcucci. "Per una coscienza vitruviana. Registo cronologico e critico delle edizioni, delle traduzioni e delle ricerche più important sul trattato Latino *De architectura Libri X* di Marco Vitruvio Pollione." *Studi e Documenti di Architettura*, no. 8 (September 1978): 11–184.

Valleriani, Matteo. *Galileo Engineer.* Dordrecht: Springer, 2010.

Veltman, Kim H., and Kenneth D. Keele. *Linear Perspective and the Visual Dimensions of Science and Art.* Munich: Deutscher Kunstverlag, [?1986].

Verna, Catherine. *Les mines et les forges des Cisterciens en Champagne méridionale et en Bourgogne du nord, XIIe–XVe siècle.* Paris: Association pour l'Édition et la Diffusion del Études Historiques, 1995.

Vesalius, Andreas. *Andreae Vesalii Tabulae anatomicae sex.* Venice: B. Vitalis, 1538.

———. *De humani corporis fabrica.* Translated by Daniel H. Garrison and Malcolm Hast (in progress). http://vesalius.northwestern.edu.

———. *De humani corporis fabrica libri septem.* Basil: [Ex officina I. Oporini, 1543].

———. *On the Fabric of the Human Body: A Translation of "De humani corporis fabrica libris septem."* 7 vols. Translated by William Frank Richardson and John Burd Carman. San Francisco: Norman, 1998–2009.

Vickers, Brian. "The 'New Historiography' and the Limits of Alchemy." *Annals of Science* 65 (January 2008): 127–156.

Vitruvius. *De architectura libri dece tr. de latino in vulgare, affigurati: commentate: & conmirando ordine insigniti [da Caesare Caesariano].* [Como]: G. da Ponte, 1521.

———. *On Architecture.* Edited and translated by Frank Granger. 2 vols. Loeb. Cambridge, Mass.: Harvard University Press, 1931–1934.

———. *Ten Books on Architecture.* Edited by Ingrid D. Rowland and Thomas Noble Howe. Cambridge: Cambridge University Press, 1999.

———. *Ten Books on Architecture: The Corsini Incunabulum with the Annotations and Autograph Drawings of Giovanni Battista da Sangallo.* Edited by Ingrid D. Rowland. Rome: Edizioni dell'Elefante, 2003.

von Staden, Heinrich. "Physis and Technē in Greek Medicine." In *The Artificial and the Natural,* edited by Bensaude-Vincent and Newman, 21–49.

Walker, Paul Robert. *The Feud That Sparked the Renaissance: How Brunelleschi and Ghiberti Changed the Art World.* New York: HarperCollins, 2002.

Wallis, Faith. "Michael of Rhodes and Time Reckoning: Calendar, Almanac, Prognostication." In *Book of Michael of Rhodes,* edited by Long, McGee, and Stahl, 3:281–319.

Walton, Steven A. "State Building through Building for the State: Foreign and Domestic Expertise in Tudor Fortification." In *Expertise: Practical Knowledge and the Early Modern State,* edited by Eric H. Ash. *Osiris,* 2d ser., 25 (2010): 66–84.

Wear, Andrew. "Medicine in Early Modern Europe, 1500–1700." In *The Western Medical Tradition, 800 BC to AD 1800,* edited by Lawrence C. Conrad, Michael Neve, Vivian Nutton, Roy Porter, and Andrew Wear, 207–361. Cambridge: Cambridge University Press, 1995.

Westermann, Angelika. *Entwicklungsprobleme der Vorderösterreichischen Montanwirtschaft im 16. Jahrhundert: Eine verwaltungs-, rechts-, wirtschafts-, und sozialgeschichtliche Studie als Vobereitung für einen multiperspektivischen Geschichtsunterricht.* Idstein: Schulz-Kirchner, 1993.

Westfall, Carroll William. *In This Most Perfect Paradise: Alberti, Nicholas V, and the Invention of Conscious Urban Planning in Rome, 1447–55.* University Park: Pennsylvania State University Press, 1974.

Westman, Robert S. *The Copernican Question: Prognostication, Skepticism, and the Celestial Order.* Berkeley: University of California Press, 2011.

Whitney, Elspeth. *Paradise Restored: The Mechanical Arts from Antiquity through the Thirteenth Century.* Transactions of the American Philosophical Society, n.s. 80, pt. 1. (1990).

Wiggershaus, Rolf. *The Frankfurt School: Its History, Theories, and Political Significance.* Translated by Michael Robertson. Cambridge, Mass.: MIT Press, 1994.

Willmoth, Frances. *Sir Jonas Moore: Practical Mathematics and Restorations Science.* Woodbridge, Suffolk: Boydell Press, 1993.

Wilson, N. G. "Vettor Fausto, Professor of Greek and Naval Architect." In *The Uses of Greek and Latin,* edited by A. C. Dionisotti, Anthony Grafton, and Jill Kraye, 89–95. London: Warburg Institute, 1988.

Witt, Ronald G. *In the Footsteps of the Ancients: The Origins of Humanism from Lovato to Bruni.* Leiden: Brill, 2000.

Wulz, Monica. "Collective Cognitive Processes around 1930: Edgar Zilsel's Epistemology of Mass Phenomena." http://philsci-archive.pitt.edu/archive/00004740/.

Wundram, Manfred. "(1) Lorenzo (di Cione) Ghiberti." In *Dictionary of Art,* edited by Turner, 12:536–545.

Zilsel, Edgar. "Appendix II: Laws of Nature and Historical Laws." In Zilsel, *Social Origins of Modern Science,* edited by Raven, Krohn, and Cohen, 233–234.

———. *Die Entstehung des Geniebegriffes: Ein Beitrag zur Ideengeschichte der Antike und des Frühkapitalismus.* Preface by H. Maus. 1926. Hildesheim: Olms Verlag, 1972.

———. "The Genesis of the Concept of Physical Law." *Philosophical Review* 51 (May 1942): 245–279.

———. "The Genesis of the Concept of Scientific Progress." *Journal of the History of Ideas* 6 (June 1945): 325–349.

———. "The Methods of Humanism." In Zilsel, *Social Origins of Modern Science,* edited by Raven, Krohn, and Cohen, 22–64.

———. "The Origins of Gilbert's Scientific Method." *Journal of the History of Ideas* 2 (January 1941): 1–32.

———. "Physics and the Problem of Historico-sociological Laws." In Zilsel, *The Social Origins of Modern Science,* edited by Raven, Krohn, and Cohen, 200–213.

———. "Problems of Empiricism." In *The Development of Rationalism and Empiricism. International Encyclopedia of United Science,* vol. 2, no. 8, edited by Otto Neurath, 53–94. Chicago: University of Chicago Press, 1941.

———. *The Social Origins of Modern Science.* Edited by Diederick Raven, Wolfgang Krohn, and Robert S. Cohen. Dordrecht: Kluwer Academic, 2000.

———. "The Sociological Roots of Science." *American Journal of Sociology* 47 (January 1942): 544–562.

———. *Die sozialen Ursprünge der neuzeitlichen Wissenschaft.* Edited and translated by Wolfgang Krohn. Biobibliographical notes by Jörn Behrmann. Frankfurt am Main: Suhrkamp, 1976.

Zilsel, Paul. "Portrait of My Father." *Shmate* 1 (April/May 1982): 12–13.

Zolo, Danilo. *Reflexive Epistemology: The Philosophical Legacy of Otto Neurath.* Translated by David McKie. Dordrecht: Kluwer Academic, 1989.

Zwijnenberg, Robert. *The Writings and Drawings of Leonardo da Vinci: Order and Chaos in Early Modern Thought.* Translated by Caroline A. van Eck. Cambridge: Cambridge University Press, 1999.

Index

Page numbers in bold refer to illustrations

Duomo (dome of Florentine cathedral). *See* Santa Maria del Fiore

Durkheim, Emile, 23

Eamon, William, 27

Egypt, 116, 119. *See also* obelisk(s)

empirical values and methodologies, 1–2, 3, 7, 9, 21, 23–24, 30, 33–37, 67–69, 96, 125, 126, 127–31

Engels, Friedrich, 11, 17

engineer. *See* architect/engineer

engineering, 41–47, 71–73, 94, 96, 112, 116–20, 131: hydraulic, 8, 63, 83, 84, **86,** 112–16; modern, 94, 113

England, 98, 99–100, 120, 121

epigraphy, 81, 84

Ercker, Lazarus, 111

Erzgebirge, 108, 111

Este: Alfonso d', 123; Ecole I, 83, 87

experience, 7–8, 12–13, 33–37, 111, 125, 130

experimentation, 3, 7–8, 12–13, 33–37, 96, 105, 109–10, 124

exploration, oceanic, 2–3

fabrica and *ratiocinatio*, 63, 66, 72, 76, 77–78, 80–81, 88

Fausto, Vettor, 105–6

fencing, 116

Ferrara, 125. See also under *De architectura* by Vitruvius

Filarete (Antonio Averlino), 76–80, **79,** 92, 126

Filelfo (Francesco da Tolentino), 77

Flanders, 97

flooding. *See* engineering, hydraulic

Fontana, Domenico, **117,** 119–20

fortification and fortresses, 37, 44, 80, 84, 85, 96, 99, 100, 106, 123, 125

fountains, 81, 83, 85, 113, 121

France, 84–85, 98, 108. See also Paris

Francesco di Giorgio, 41–47, **42, 43, 45,** 80–81, 83–84, 92, 126

Frankfurt, 6–7, 17–18, 19, 20, 21. *See also* Institute for Social Research

Frontinus, 85

Galen, 34, 58, 86

Galilei, Galileo, 1, 22, 23, 24, 50, 106, 130, 711

Galison, Peter, 94, 125

Gallo, Agostino, 91

gardens, 37, 39, 40, 41, 83, 121

gears, 43–44, 46, 48–50, **49**

Gerlach, Kurt, 17

Germany and the German states, 6, 14, 15, 20, 35, 98, 108, 111, 120

Ghiberti, Lorenzo, 23, 70, 73–76, **75,** 92

gift exchange, 94

Giocondo, Giovanni, 83–85, **86,** 92, 106, 126

Giovanni da Porlezza, 120

goldsmiths and goldsmithing, 23, 33, 66, 73–74, **75,** 110, 129

Golinski, Jan, 26

Graham, Loren, 15–16

Grant, Edward, 5

Greek, 30, 39, 53, 74, 77, 106, 115

Gregory XIII, pope, 116, 119

Grosseteste, Robert, 34

Grossmann, Henryk, 6–7, 10, 18–20, 21, 22, 127: criticism of Franz Borkenau, 18–19

Grünberg, Carl, 17

gun carriages, 97, 98

gunpowder artillery, 37, 84, 96–97, 101, 106–7, 108, **112,** 124, 125, 129. *See also* caliber of guns; weapons and war

Hall, A. Rupert, 25, 127

Harkness, Deborah, 28

Hart, Vaughan, 51

Hartmann, Georg, 98–99, 106

Henry VIII, king of England, 98, 99

Hessen, Boris, 6, 10, 15–17, 20, 21, 22, 23, 127

Hicks, Peter, 51

hieroglyphs, 119

Hippocrates, 31, 34, 86

historians of science, 7, 23, 24–25, 26–28, 34–35, 127

history of science, 4, 10–29: and
experience and experiment, 12–13,
24, 33–37; and theory, 24–25, 127;
constructivism in, 26–28
Hobbes, Thomas, 27
Hogge, Ralph of Buxted, 99
Hooke, Robert, 35, **36**
Horkheimer, Max, 19, 20
Hoyningen-Huene, Paul, 26
humanism, 5–6, 21, 62, 92, 114,
120–21, 123, 129: and Alberti, 69–70,
73; and Antonio Manetti, 66; and
Filelfo, 77; and Francesco di Giogio,
41, 47; and Pomponio Leto, 81–82;
and Vettor Fausto, 105–6; and
Vitruvian studies, 82–83, 85–86
Hungary, 17–18, 108
Hypnerotomachia Poliphili, 38-41,**40**

idealism, 11, 24–25
illustrations, 111, 112, 119: of machines,
36, 42, 43, 45, 49, 84, **90, 109,**
111, 112; of the *De architectura* by
Vitruvius, **65,** 84, 85, 86, **86,** 89, **90,**
122
innovation, technical, 96–101, 109–10
Innsbruck, 97–98, **98**
inscriptions. *See* epigraphy
Institute for Social Research
(Frankfurt), 6–7, 17–19, 20–21
instruments and instrument makers,
3, 37, 84, 93, 94, 96, 98–99, 107, 121,
129
investigation of nature, 1–6, 9, 30,
33–35, 46–47, 50, 56–60, 114, 116,
126, 127, 129, 130

Jardine, Lisa, 37
John VIII Palaiologos, Byzantine
emperor, 104–5
Julius II, pope, 85
Julius, duke of Braunschweig-
Wolfenbüttel, 123–24

Kemp, Martin, 70
Kepler, Johannes, 130
Kolman, Arnošt, 15

Koyré, Alexandre, 24–25, 127
Kristeller, Paul, 31
Kuhn, Thomas S., 26–27

language, vernacular, 4, 22–23, 69,
70, 92, 126, 129
Lässl, Ludwig, 111
Latin, 4, 8, 22–23, 39, 61, 62, 66,
74, 80, 81, 82, 87, 92, 105: and
humanism, 5–6, 65, 69, 70–71,
77, 86, 95, 111, 126; learned by
workshop-trained individuals, 62,
65, 66, 74, 80
Latour, Bruno, 27
Leo X, pope, 85
Leonardo da Vinci, 20, 22, 23, 47–50,
93, 126
Leto, Pomponio, 81, 82, 84
Leviathan and the Air-Pump, The by
Steven Shapin and Simon Schaffer,
27
Löffler, Gregor, 97
logical empiricism, 6, 10, 12–13
Luca della Robbia, 70–71
Lucretius, 64
Lukács, Georg, 17–18

machines, 8, 11, 65, 67, 69, 91, 92,
93, 108, 127, 129: and Francesco
di Giorgio, 41–50; Hessen and
Grossmann's view of, 16–17, 19–20,
22
Maiano, Giuliano da, 83
Manetti, Antonio, 66–69
Manutius, Aldus, 38
maps, 113, 115, **115,** 123
Marx, Karl, 11, 17
Marxism and Marxists, 6–7, 10,
11–22, 23, 127, 129. *See also* Austro-
Marxism, anti-Marxists
Masaccio, 71
Masini, Francesco, 117–18
mathematics, 12, 13, 23, 69, 85, 87,
99–100, 101, 103, 104–5, 106–7, 130
Mauss, Marcel, 95
Maximilian I, emperor, 97

McGee, David, 104
measurement, 3, 8, 37, 67, 69, 78, 84, 93, 96, 99, 121, **122,** 129, 130
mechanical arts, 10, 30, 31, 37, 100, 127. *See also* practical and technical arts
mechanical world view, 10, 18, 19, 22
mechanics, 16, 19–20, 31
Medici: Giuliano de, 85; Lorenzo de, 71, 84; Piero de, 77
medicine, 31, 34, 56, 58, 60, 61, 63, 76, 86. *See also* anatomy; Vesalius
Mercati, Michele, 116, 119
Merton, Robert K., 7, 10, 11, 23–24, 25, 127
metals and metallurgy, 33, 97, 107–12
Michael of Rhodes, 101–105, **103, 104,** 106, 126
Milan, 76, 87, 91, 116
mills, **42, 43,** 43–47, **45**
mining and ore processing, 2, 8, 23–24, 33, 93, 96, 97, 107–12, **109,** 120, 123–24, 125
minting, 33, 107, 110
models, 4, 105
Montefeltre, Federico I, 81
More, Jonas, 100

Naples, 83–85
natural history, 3, 28, 37, 95
natural philosophy, 2, 7, 8, 61, 116
nature and the natural world, 7, 30–61. *See also* art and nature; investigation of nature
navigation, 101, 104, 116
Nazis, 14–15, 20
Neurath, Otto, 13
new sciences, 1, 2, 3, 10, 23–24, 28.130
Newman, William R., 31, 32
Newton, Isaac, 1, 15–16, 24, 33, 130
Norman, Robert, 21–22
numismatics, 5, 62, 81
Nummedal, Tara, 28, 33
Nuremberg, 96, 98, 110

obelisk(s), 113, 116–20, **118**
objects, cultural significance of, 3, 37, 38–41, 60, 88, 90
observation, 2, 3, 30, 33, 34, 35, 37, 111, 129, 130: of buildings, 53–56, 67–68; of machines, 19–20, 50
Ogilvie, Brian, 28
Olschki, Leonardo, 7, 10, 22–23
optics, 34, 61, 74

Pacioli, Luca, 23
painting, 3, 4, 30–31, 37, 51, 60, 63, 66, 120, 121, 128, 130, 131: and Alberti, 6, 69, 70; and Cesariano, 87; and Filarete, 78, 80; and Ghiberti, 74, 76. *See also* perspective, artist's
palaces and villas, 3, 37, 40, 82, 83, 85, 86, 87, 95, 113, 121, 125, 127
Palissy, Bernard, 7, 25, 35, **36**
Palladio, 120–23, 126
Palopano, Nicolò, 101, 105
Paris, 20, 24, 85, 121
Park, Katharine, 34–35
Parsons, Talcott, 23
patrons and patronage, 4, 6, 33, 41, 44, 76–77, 95, 96, 101, 103, 113, 120, 121, 123
Paul of Taranto, 32–33
Peckham, John, 74
perspective, artist's, 3, 30–31, 66, 70, 71, 74, 121
Perugia, 123
Peto, Luca, 115–16
Petrarch, 5
Philip II, king of Spain, 99
physics, twentieth century, 12, 13, 15–16, 94
physis. See nature
Piccolpasso, Cipriano, 123, **124**
Piero della Francesca, 23
Pigafetta, Filippo, 119
Pirova, Aloisio, 91
Pisano, Andrea, 73
Pittoni, Girolamo, 120
Pius IV, pope, 114, 115

terminology, technical, 85, 112, 125–26
theater, 51, 80, 82
theory, role in science, 24–25
Tiber River, 113–16
Titian, 51
trading zones, 8, 93, 94–126, 128, 131
treatises. *See* books and publication; writings
Trevisi, Antonio, 114–15, 116
Trissino, Gian Giorgio, 120–21, 123

unity of knowledge, 13, 14, 135 n8
universities, 2, 4–5, 6, 61, 78, 81, 121, 129
university-trained individuals, 129: and relationship to skilled individuals, 8, 51, 56, 62, 65–66, 125. *See also* humanism

Valleriani, Matteo, 106
vaults, 67
Venice, 85, 96, 100–103, **102**
Verrocchio, Andrea del, 47
Vesalius, Andrea, 50–60, **52, 55, 57, 58**
Vicenza, 120
Vienna, 6, 12, 13, 14, 17, 19, 21: adult education in, 13, 21
Vienna Circle, 6, 12, 13, 14, 20–21
villas. *See* palaces and villas
Visconti, Antonio
visual arts and culture, 3, 57–60, 71, 94, 113. *See also* painting; perspective, artist's
Vitruvian man, 93
Vitruvius and the Vitruvian tradition, 8, 41, 44, 61, 62–93, 77–78, 80–93, 95. *See also* De architectura of Vitruvius

weapons and war, 2, 23–24, 37, 72, 80, 97, 116, 120. *See also* gunpowder artillery
Weber, Max, 23
weights and measures, 115, 119
Weil, Felix, 17
Weil, Hermann, 17, 20
Witelo, 74
workshops, craft, 8
writings, autobiographical, 87–88, **89,** 101
writings, practical and technical, 6, 7, 23, 37, 96, 97, 106–7, 125: by Michael of Rhodes, 101–5; on anatomy, 50–61; on architecture, 6, 8, 50, 53–56, **54,** 62–93, **79,** 121, **122**; on ceramics, 123; on engineering, 41–47, 113–20, 116, 121; on machines, 48–50, 63, 64, 71, 72, 74, 123–24; on mines and metallurgy, 106, 110–12; on ships and shipbuilding, 101, **104,** 124
Wundrum, Manfred, 74

Zaneto, Giovanni di, 106
Zilsel, Edgar, 6, 10, 12, 13–15, **14,** 20–21, 92, 127, 129
zodiac, 101–3